Environmental Control in Petroleum Engineering

Environmental Control in Petroleum Engineering

Editor

Srajan Tripathi

scitus
academics

Environmental Control in Petroleum Engineering
Edited by **Srajan Tripathi**

Printed in 2017

ISBN: 978-1-68117-366-5

Library of Congress Control Number: 2015941554

SCITUS Academics LLC,
616, Corporate Way, Suite 2, 4766,
Valley Cottage, NY 10989

www.scitusacademics.com

Notice

Contents

Preface

Environmental Control in Petroleum Engineering is essential for industry personnel with little or no training in environmental issues as well as petroleum engineering students. The petroleum industry must minimize the environmental impact of its various operations. This extensively researched book assembles a tremendous amount of practical information to help reduce and control the environmental consequences of producing and processing petroleum and natural gas. The best way to treat pollution is not to create it in the first place. This book shows you how to plan and manage production activities to minimize and even eliminate some environmental problems without severely disrupting operations. It focuses on ways to treat drilling and production wastes to reduce toxicity and/or volume before their ultimate disposal. Other topics include United States federal environmental regulations, sensitive habitats, major U.S. chemical waste exchanges, and offshore releases of oil.

Editor

Chapter 1

Thermolysis of High-Density Polyethylene to Petroleum Products

Sachin Kumar and R. K. Singh

Department of Chemical Engineering, National Institute of Technology Rourkela, Orissa 769008, India

ABSTRACT

Thermal degradation of plastic polymers is becoming an increasingly important method for the conversion of plastic materials into valuable chemicals and oil products. In this work, virgin high-density polyethylene (HDPE) was chosen as a material for pyrolysis. A simple pyrolysis reactor system has been used to pyrolyse virgin

HDPE with an objective to optimize the liquid product yield at a temperature range of 400°C to 550°C. The chemical analysis of the HDPE pyrolytic oil showed the presence of functional groups such as alkanes, alkenes, alcohols, ethers, carboxylic acids, esters, and phenyl ring substitution bands. The composition of the pyrolytic oil was analyzed using GC-MS, and it was found that the main constituents were n-Octadecane, n-Heptadecane, 1-Pentadecene, Octadecane, Pentadecane, and 1-Nonadecene. The physical properties of the obtained pyrolytic oil were close to those of mixture of petroleum products.

INTRODUCTION

Plastic materials comprise a steadily increasing proportion of the municipal and industrial waste going into landfill. Owing to the huge amount of plastic wastes and environmental pressures, recycling of plastics has become a predominant subject in today's plastics industry. Development of technologies for reducing plastic waste, which are acceptable from the environmental standpoint and are cost-effective, has proven to be a difficult challenge because of the complexities inherent in the reuse of polymers. Establishing optimal processes for the reuse/recycling of plastic materials, thus, remains a worldwide challenge in the new century. Plastic materials find applications in agriculture as well as in plastic packaging, which is a high-volume market owing to the many advantages of plastics over other traditional materials. However, such materials are also the most visible in the waste stream and have received a great deal of public criticism as solid materials have comparatively short life-cycles and usually are nondegradable.

Thermal cracking, or pyrolysis, involves the degradation of the polymeric materials by heating in the absence of oxygen. The process is usually conducted at temperatures between 500 and 800°C and results in the formation of a carbonized char and a volatile fraction that may be separated into condensable hydrocarbon oil and a noncondensable high calorific value gas. The proportion of each fraction and its precise composition depend primarily on the nature of the plastic waste and on process conditions as well.

In pyrolytic processes, a proportion of species generated directly from the initial degradation reaction are transformed into secondary products due to the occurrence of inter- and intramolecular reactions. The extent and the nature of these reactions depend both on the reaction temperature and also on the residence of the products in the reaction zone, an aspect that is primarily affected by the reactor design.

In addition, reactor design also plays a fundamental role, as it has to overcome problems related to the low thermal conductivity and high viscosity of the molten polymers. Several types of reactors have been reported in the literature, the most frequent being fluidized bed reactors, batch reactors, and screw kiln reactors [2].

Characteristics of thermal degradation of heavy hydrocarbons can be described with the following items.

- High production of C_1s and C_2s in the gas product.
- Olefins are less branched.
- Some diolefins made at high temperature.
- Gasoline selectivity is poor; that is, oil products have a wide distribution of molecular weight.
- Gas and coke products are high.
- Reactions are slow compared with catalytic reactions.

High-density polyethylene (HDPE) is the third-largest commodity plastic material in the world, after polyvinyl chloride and polypropylene in terms of volume. It is a thermoplastic material composed of carbon and hydrogen atoms joined together forming high-molecular-weight products. The effect of temperature and the type of reactor on the pyrolysis of HDPE has been studied, and some of the results are reviewed.

Wallis and Bhatia have done the thermal degradation of high-density polyethylene in a reactive extruder at various screw speeds with reaction temperatures of 400°C and 425°C. A continuous kinetic model was used to describe the degradation of the high-density polyethylene in the reactive extruder. It was found that purely random breakage and a scission rate which had a power-law dependence on molecular size of 0.474 best described the experimental data. The greatest discrepancy between the model prediction and the experimental data was the large molecular size region at short residence times; however, this only accounted for a very small percentage of the total distribution and was attributed to the presence of fast initiation reaction mechanism that was only significant at low conversions [3].

Conesa et al. studied the production of gases from polyethylene (HDPE) at five nominal temperatures (ranging from 500°C to 900°C) using a fluidized sand bed reactor. HDPE primary decomposition and wax cracking reactions take place inside the reactor. Yields of 13 pyrolysis products (methane, ethane, ethylene, propane, propylene, acetylene, butane, butylenes, pentane, benzene, toluene, xylenes, and styrene) were analyzed as a function of the operating conditions. From the study of HDPE pyrolysis in a fluidized sand bed reactor, they have found that the yield of total gas obtained increases in the range 500°C–800°C from 5.7 to 94.5%; at higher temperatures, the yield of total gas decreases slightly; the formation of methane, benzene, and toluene is favored by high residence times, but ethane, ethylene, propane, propylene, butane, butylenes, and pentane undergo cracking to different extents at increasing residence times and/or temperature; and the maximum yield of total gas obtained at 800°C from HDPE pyrolysis is 94.5% with the following composition: 20% methane, 3.8% ethane, 37% ethylene, 0.2% propane, 4.7% propylene, 0.3% butane, 0.4% butylenes, 2.2% pentane, 24% benzene, 2.1% toluene, 0.01% acetylene, and 0.02% xylenes and styrene [4].

Walendziewski and Steininger reported the thermal degradation of polyethylene in the temperature range 370–450°C. In the case of thermal degradation of polyethylene, an increase in degradation temperature led to an increase of gas and liquid products, but a decrease of residue (boiling point > 360°C). However, the increase of gas was not too large as compared to the sharp decrease of residue with increase of temperature. The result of analysis of gas products obtained by the pyrolysis of polyethylene at 400°C is summarized in Table 1 [1].

Table 1: Composition of gas products obtained from pyrolysis of polyethylene at 400°C [1]

Component	Thermal	Catalytic	Hydrocracking
Methane	22.7	12.4	21.1
Ethane	27.4	20.4	21.2
Ethylene	1.4	2.3	0.1
C_3	26.6	30.4	23.7
C_4	11.0	20.3	20.7
C_5	6.9	5.6	7.3
C_6	2.1	3.3	3.8

Walendziewski carried out two series of experiments of waste polymers cracking. The first series of polymer cracking experiments was carried out in glass reactor of 0.5 dm^3 volume at atmospheric pressure and in a temperature range 350–420°C, the second one in autoclaves under hydrogen pressure (≈3–5 MPa) in the temperature range 380–440°C. The influence of cracking parameters, that is, reaction temperature, presence and amount of cracking catalysts, composition of the polymer feed on product yields, and composition of gas and liquid fractions is discussed. It was stated that the proper selection of the process parameters makes it possible to control, in the limited range, the product composition distribution as well as yields and composition of gas, gasoline, and diesel oil fractions [5].

Walendziewski carried out the experiments of waste polymers cracking in a continuous-flow tube reactor. The main components of the reactor unit were a screw extruder as a waste plastics feeder and a tube cracking reactor equipped with an internal screw mixer. Cracking process was realized at the temperature range 420–480°C and raw material feeding rate from 0.3 up to 2.4 kg/h. The principal process products, gaseous and liquid hydrocarbon fractions, are similar to the refinery cracking products. They are unstable due to their high olefins content (especially from polystyrene cracking), and their chemical composition and properties strongly depend on the applied feed composition, that is, shares of polyethylene, polypropylene, and polystyrene. The material balance experiments showed that the main products, liquid or solid materials in ambient temperature, contain typically 20–40% of gasoline fractions (range of boiling point 35–180°C) and 60–80% of light gas oil fractions (initial boiling point > 180°C). The solid carbon residues are similar to coal cokes and even contain 50% mineral components. Their calorific values attain 20 MJ/kg and they are solid products of quality similar to brown coals [6].

A number of studies have been reported in which a range of catalysts and reaction conditions have been employed to convert waste plastics into the hydrocarbon liquid using pyrolysis during the past four decades. The most commonly used catalysts in the catalytic degradation of high-density polyethylene are solid acids (zeolite, silica-alumina) [7–13] and spent FCC [14, 15].

This work focuses on characterization of liquid product obtained from thermal pyrolysis of virgin high-density polyethylene at different temperature ranges. Thermal pyrolysis of high-density polyethylene pellets was done in a semibatch reactor at a temperature range of 400°C to 550°C and at a heating rate of 20°C/min. The effect of pyrolytic temperature on reaction time, liquid yield, and volatiles was also studied. The obtained liquid product was characterized by different physical and chemical properties using GC-MS and FTIR.

MATERIALS AND METHODS

HDPE pellets (2.5 mm in size) obtained from Reliance Industries Ltd., India, with density 0.945 g/cc^3, Melt Flow Index (MFI) value 0.2–15 g/10 min^{-1} (at 190°C and 2.16 kg load), and melting point 133°C were used for experiments. These plastic pellets were used directly in the thermal pyrolysis reaction. The proximate analysis of HDPE pellets was done by ASTM D3173-75 and ultimate analysis was done using CHNS analyzer (Elementar Vario El Cube CHNSO). Calorific value of the raw material was found by ASTM D5868-10a.

Thermogravimetric analysis of the HDPE sample was carried out with a Shimadzu DTG-60/60H instrument. A known weight of the sample was heated in a silica crucible at a constant heating rate of 293 K/min operating in a stream of nitrogen with a flow rate of 40 mL/min from 32°C to 700°C.

The pyrolysis setup consists of a semibatch reactor made of stainless-steel tube (length: 145 mm, internal diameter: 37 mm, and outer diameter: 41 mm) sealed at one end and an outlet tube at other end as shown in the previous study [16]. The reactor is heated externally by an electric furnace, with the temperature being measured by a Cr-Al : K-type thermocouple fixed inside the reactor, and temperature is controlled by external PID controller. 20 g of HDPE sample was loaded in each pyrolysis reaction. The condensable liquid products/wax were collected through the condenser and weighed. After pyrolysis, the solid residue (wax) left out inside the reactor was weighed. Then, the weight of gaseous/volatile product was calculated from the material balance. Reactions were carried out at different temperatures ranging from 400 to 550°C.

Fourier transform infrared spectroscopy (FTIR) of the pyrolysis oil obtained at optimum condition was taken in a Perkin-Elmer Fourier transformed infrared spectrophotometer with resolution of 4 cm^{-1}, in the range of 400–4000 cm^{-1} using Nujol mull as reference to know the functional group composition. The components of liquid product were analyzed using GC-MS-QP 2010 (Shimadzu) using flame ionization detector. The GC conditions, column oven temperature progress, column used, and MS conditions are given in the Table 2.

Table 2: GC-MS conditions

	GC-MS-OP 2010 Shimadzu	
GC conditions		
Column oven temperature	70°C	
Injection mode	Split	
Injection temperature	200°C	
Split ratio	10	

Flow control mode	Linear velocity	
Column flow	1.51 mL/min	
Carrier gas	Helium 99.9995% purity	
Column oven temperature progress		
Rate	Temperature (°C)	Hold time (min)
—	70	2
10	300	7.0 (32 min total)
Column: DB-5		
Length	30.0 m	
Diameter	0.25 mm	
Film thickness	0.25 μm	
MS conditions		
Ion source temperature	200°C	
Interface temperature	240°C	
Start	40	
End	1000	

RESULT AND DISCUSSION

Proximate and Ultimate Analyses of Virgin HDPE

The proximate and ultimate analyses of virgin HDPE sample are shown in Table 3. The volatile matter is 99.9% in the proximate analysis, due to the negligible percentage of ash in virgin HDPE sample; its degradation occurs with minimal formation of residue. The oxygen is 2.51% in the ultimate analysis of virgin HDPE. The nitrogen and oxygen in the virgin HDPE sample may not be due to the fillers but due to the other ingredients which are added to resin during the manufacturing of HDPE.

Table 3: Proximate and ultimate analyses of virgin HDPE.

Properties	Virgin HDPE
Proximate analysis	
Moisture content	0.00
Volatile matter	99.92
Fixed carbon	0.00
Ash content	0.08
Ultimate analysis	
Carbon (C)	83.29
Hydrogen (H)	13.93
Nitrogen (N)	0.20
Sulphur (S)	0.07
Oxygen (O)/others	2.51
GCV (MJ/kg)	47.64

TGA and DTG Analyses of Virgin HDPE Sample

Thermogravimetric analysis (TGA) is a thermal analysis technique which measures the weight change in a material as a function of temperature and time in a controlled environment. This can be very useful to investigate the thermal stability of a material or to investigate its behavior in different atmospheres (e.g., inert or oxidizing). TGA is applied in determination of the study of thermal stability/degradation of virgin HDPE in various ranges of temperature.

From the TGA curve as shown in Figure 1, the virgin HDPE degradation started at 380°C and was completed at 510°C for a heating rate of 293 K/min in nitrogen atmosphere. The degradation temperature at which weight loss of 50% (T_{50}) takes place was about 460°C for virgin HDPE. The temperature range for waste

HDPE was 390°C to 490°C, and maximum weight loss occurred at temperature 440°C as demonstrated [16]. A similar trend of nature during the analysis of HDPE decomposition by TGA/DTG has been reported by Aboulkas et al. [19]. Differential thermogravimetry (DTG) is exactly the same as TGA, except the mass loss versus time output is differentiated automatically to give the mass loss rate versus time. Generally, both the mass loss and mass loss rate versus time are produced automatically. This is quite convenient as the rate of thermal decomposition is proportional to the volatilization or mass loss rate. Differential thermogravimetry (DTG) curve for virgin HDPE contains only one peak; this indicates that there is only one degradation step in Figure1, showing that the dominant peak is from 390°C to 510°C where the conversion takes place. Similarly, maximum weight loss occurs at 498°C from DTG profile of HDPE pellets [20].

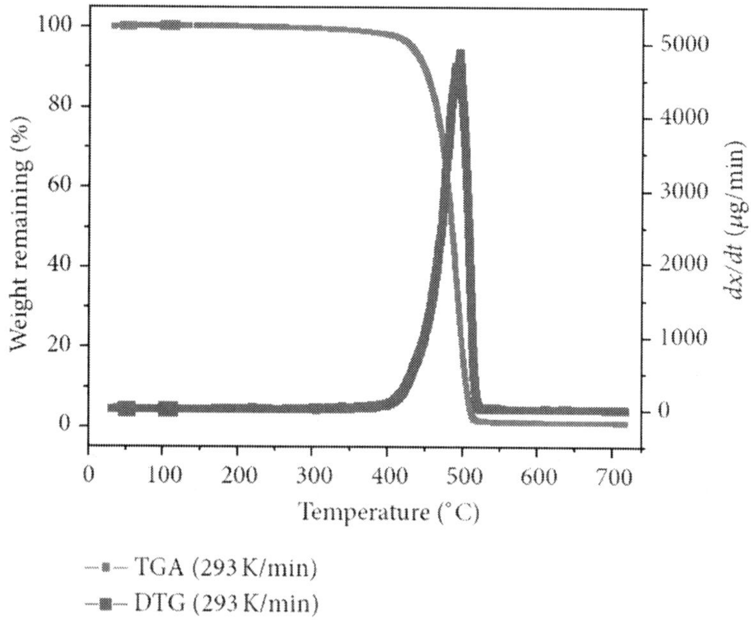

Figure 1: TGA and DTG curves of virgin HDPE.

Effect of Temperature on Product Distribution

The pyrolysis of virgin HDPE yielded four different products, that is, oil, gas, wax, and residue or coke. The distributions of these fractions are different at different temperatures and are shown in Table 4.

Table 4: Distribution of different fractions at different temperatures in thermal pyrolysis of virgin HDPE.

Temperature (°C)	Oil (wt%)	Residue (wt%)	Wax (wt%)	Gas/volatile (wt%)	Reaction time (min)
400	31.3	5.65	7.7	45.35	680
450	52.46	3.95	8.9	34.69	175
500	44.32	1.29	28.99	25.4	80
550	8.83	0.68	52.02	38.47	50

The condensable oil/wax (a mixture of alkanes that falls within the $20 \leq n \leq 40$ range; they are found in the solid state at room temperature and begin to enter the liquid phase past approximately 37°C) and the noncondensable gas/volatiles fractions of the reaction constituted major product as compared to the solid residue (coke) fractions. The condensable product obtained at low temperature (400°C) was low viscous liquids. With increase in temperature, the liquid became viscous/wax at and above 450°C. The hydrocarbon is continuously cracking; the wax may be representative of the intermediate-molecular-weight products. The recovery of condensable fraction was very low at low temperature that is, at 400°C and increased with gradual increase of temperature. From Table 4, it is observed that at low temperature the reaction time was more, due to which secondary cracking of the pyrolysis product occurred inside the reactor and resulted in highly volatile product. The low-temperature molecular weight changes without volatilization are principally due to the scission of weak links, such as oxygen, incorporated into the main chain as impurities. Similarly, the low liquid yield at high temperature was due to volatilizing higher-molecular-weight products before undergoing further cracking and more noncondensable gaseous/ volatile fractions due to rigorous cracking.

Effect of Temperature on Reaction Time

The effect of temperature on the reaction time of the reaction for the pyrolysis of virgin HDPE plastic is shown in Figure 2. The pyrolysis reaction rate increased and reaction time decreased with increase in temperature. High temperature supports the easy cleavage of bond and thus speeds up the reaction and lowers the reaction time. HDPE with long linear polymer chain with low branching and high degree of crystallinity led to high strength properties and thus required more time for decomposition. This shows that temperature has significant effect on reaction time and yield of liquid, wax, gaseous products, and solid residue (coke). A similar effect of temperature on reaction time for waste HDPE has been demonstrated [16].

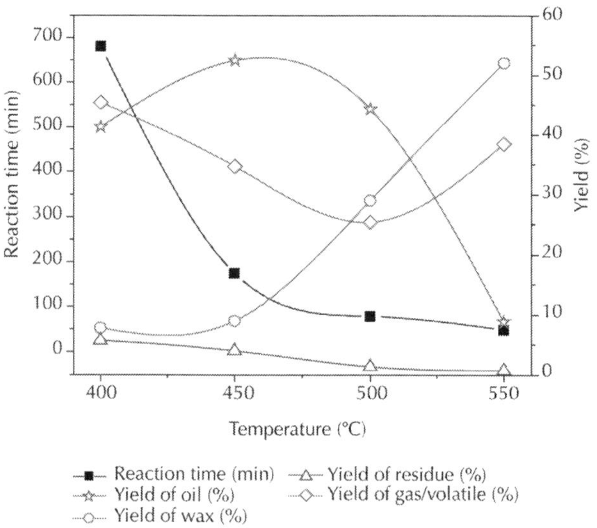

Figure 2: Effect of temperature on reaction time and product distribution.

CHARACTERIZATION OF THE LIQUID PRODUCT

FT-IR of the Oil Sample Obtained at 450°C

Fourier transform infrared spectroscopy (FTIR) is an important analysis technique which detects various characteristic functional groups present in oil. On interaction of an infrared light with oil, chemical bond will stretch, contract, and absorb infrared radiation in a specific wave length range regardless of the structure of the rest of the molecules. Figure 3 shows the FTIR spectra of virgin HDPE oil. The C–H stretching vibrations at frequency 3078.11 cm^{-1} indicate the presence of alkenes. The presence of alkanes is detected at 2918.68 $cm^{-1}$with C–H stretching vibrations. The C=C stretching vibrations at frequency 1647.39 cm^{-1} indicate the presence of alkenes. The presence of alkanes was detected by C–H scissoring and bending vibrations at 1440.22 cm^{-1}. The presence of alcohols, ethers, carboxylic acids, and esters is detected by C–O stretching vibrations at 907.61 cm^{-1}, and the C–H bending vibrations at frequency 719.92 indicates the presence of phenyl ring substitution bands. The results were found consistent when compared with the results of GC-MS.

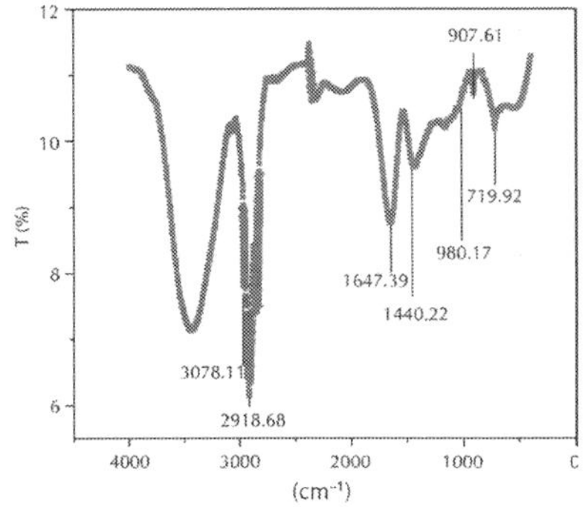

Figure 3: FT-IR spectrometry of virgin HDPE pyrolytic oil.

GC-MS of the Oil Sample

The GC-MS analysis of the oil sample obtained by the thermal pyrolysis of virgin HDPE was carried out to know the compounds present in the oil (Figure 4) and is summarized in Table 5. It has been observed that the pyrolytic oil contains around 25 compounds. Taking into account the area percentage, the highest peak areas of total ion chromatogram (TIC) of the compounds were n-Octadecane, n-Heptadecane, 1-Pentadecene, Octadecane, Pentadecane, and 1-Nonadecene. The components present in HDPE are mostly the aliphatic hydrocarbons (alkanes and alkenes) with carbon number C_{10}–C_{20}.

Table 5: GC-MS analysis of virgin HDPE pyrolytic oil

R. time (min)	Area %	Name of compound	Molecular formula
6.301	1.24	1-Decene	$C_{10}H_{20}$
6.450	1.12	Decane	$C_{10}H_{22}$
8.105	2.04	1-Undecene	$C_{11}H_{22}$
8.238	1.78	n-Undecane	$C_{11}H_{22}$
9.735	3.50	1-Dodecanol	$C_{12}H_{26}O$
9.855	3.19	n-Dodecane	$C_{12}H_{26}$
11.541	4.62	1-Tridecene	$C_{13}H_{26}$
12.615	5.30	1-Tetradecene	$C_{14}H_{28}$
12.711	4.82	Tetradecane	$C_{14}H_{30}$
12.772	0.65	7-Tetradecene	$C_{14}H_{28}$
13.909	5.40	1-Pentadecene	$C_{15}H_{30}$
13.997	5.13	Pentadecane	$C_{15}H_{32}$
15.039	0.48	1,19-Eicosadiene	$C_{20}H_{38}$
15.130	5.36	1-Hexadecene	$C_{16}H_{32}$
15.210	5.60	n-Octadecane	$C_{18}H_{36}$
15.261	0.51	Cyclohexadecane	$C_{16}H_{32}$

16.203	0.49	1,19-Eicosadiene	$C_{20}H_{38}$
16.283	5.09	1-Nonadecene	$C_{19}H_{38}$
16.357	5.52	n-Heptadecane	$C_{17}H_{36}$
16.406	0.51	1-Heptadecene	$C_{17}H_{34}$
17.378	4.43	1-Octadecene	$C_{18}H_{36}$
17.447	5.47	Octadecane	$C_{18}H_{38}$
17.493	0.69	1-Octadecene	$C_{18}H_{36}$
18.419	3.26	1-Nonadecene	$C_{19}H_{38}$
18.482	4.67	Nonadecane	$C_{19}H_{40}$

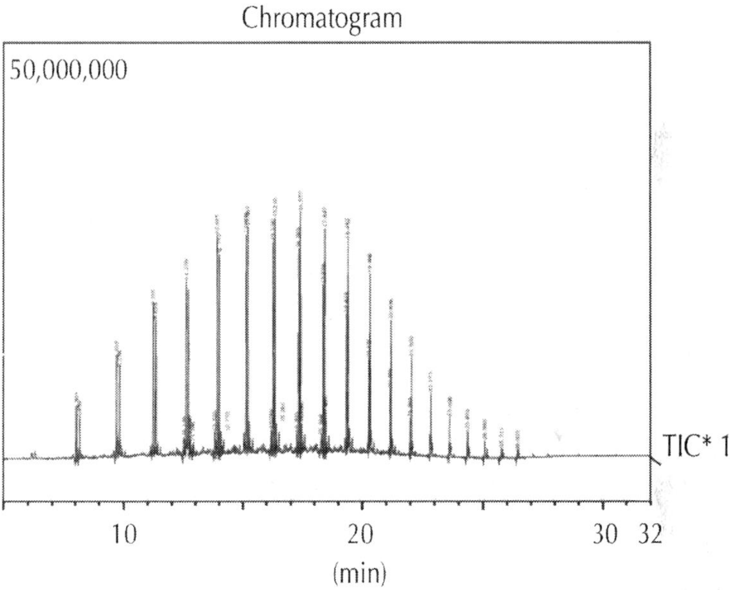

Figure 4: GC plot of oil obtained at 450°C.

Physical Properties of Oil Sample

Table 6 shows the results of physical property analysis of oil obtained from pyrolysis of virgin HDPE. The appearance of the oil is dark brownish free from visible sediments.

Table 6: Physical properties analysis of virgin HDPE pyrolytic oil

Tests	Results obtained	Test method
Specific gravity at 15°C/15°C	0.8013	IS:1448 P:16
Density at 15°C in kg/cc	0.8006	IS:1448 P:16
Kinematic viscosity at 40°C in Cst	3.3	IS:1448 P:25
Kinematic viscosity at 100°C in Cst	1.4	IS:1448 P:25
Conradson carbon residue	<0.01%	IS:1448 P:122
Flash point by Abel method	10°C	IS:1448 P:20
Fire point	15°C	IS:1448 P:20
Cloud point	28°C	IS:1448 P:10
Pour point	18°C	IS:1448 P:10
Gross calorific value in MJ/kg	44.27	IS:1448 P:6
Sulphur content	0.03%	IS:1448 P:33
Calculated Cetane Index (CCI)	70	IS:1448 P:9
Distillation		IS:1448 P:18
Initial boiling point	72°C	
Final boiling point	364°C	

From comparison with other transportation products as shown in Table 7, the density and viscosity of liquid product can be modified by blending it with commercial transportation products. The flash point of the liquid product is in a comparable range, and a pour point of 18°C is acceptable for most geographic regions. HDPE pyrolytic oil has GCV of 44 MJ/kg which is more as compared to that of gasoline and diesel; therefore, this liquid product would perform relatively superior in engines. From the distillation report of the oil, it is observed that the boiling range of the oil is 72–364°C, which infers the presence of mixture of different oil components such as gasoline, kerosene, and diesel in the oil. The liquid product contains substantial amount of volatiles as its initial boiling point is below 100°C. From this result, it is observed that these could be possible feedstocks for further upgrading or use of lighter compounds as a diesel product.

Table 7: Product properties comparison of HDPE pyrolytic oil with commercial transportation products

Properties	Specific gravity 15°C/15°C	Kinematic viscosity at 40°C (cst)	Flash point (°C)	Pour point (°C)	GCV (MJ/kg)	IBP (°C)	FBP (°C)	Chemical formula
HDPE pyrolytic oil	0.8013	3.3	10	18	44.27	72	364	C_{10}–C_{20}
Waste HDPE pyrolytic oil [16]	0.7835	1.63	1	−15	42.81	82	352	C_{19}–C_{24}
Gasoline [17]	0.72–0.78	—	−43	−40	42–46	27	225	C_4–C_{12}
Diesel [17]	0.82–0.85	2–5.5	53–80	−40 to −1	42–45	172	350	C_8–C_{25}
Biodiesel [17]	0.88	4–6	100–170	−3 to 19	37–40	315	350	C_{12}–C_{22}
Heavy product oil [18]	0.94–0.98	>200	90–180	—	−40	—	—	—

CONCLUSIONS

Thermal pyrolysis of virgin HDPE was performed in a semibatch reactor made up of stainless steel at temperature range from 400°C to 550°C and at a heating rate of 20°C/min. The liquid yield is highest at 450°C, highly volatile products are obtained at low temperature, and the products obtained at 500°C and 550°C are viscous liquid and wax. Reaction time decreases with increase in temperature. The functional group present in the virgin HDPE pyrolytic oil is similar to the other plastic pyrolytic oils given in several literatures. It was found that the pyrolytic oil contains around 25 types of compounds having carbon chain length in the range of C_{10}–C_{20}. The physical properties of pyrolytic oil obtained were in the range of other pyrolytic oils and moderate-quality products. It has been shown that a simple batch pyrolysis method can convert virgin HDPE to liquid hydrocarbon products with a significant yield which varies with temperature.

REFERENCES

1. J. Walendziewski and M. Steininger, "Thermal and catalytic conversion of waste polyolefines," Catalysis Today, vol. 65, no. 2–4, pp. 323–330, 2001.
2. S. Kumar, A. K. Panda, and R. K. Singh, "A review on tertiary recycling of high-density polyethylene to fuel," Resources, Conservation and Recycling, vol. 55, no. 11, pp. 893–910, 2011.
3. M. D. Wallis and S. K. Bhatia, "Thermal degradation of high density polyethylene in a reactive extruder," Polymer Degradation and Stability, vol. 92, no. 9, pp. 1721–1729, 2007.
4. J. A. Conesa, R. Font, A. Marcilla, and A. N. García, "Pyrolysis of polyethylene in a fluidized bed reactor," Energy & Fuels, vol. 8, no. 6, pp. 1238–1246, 1994.
5. J. Walendziewski, "Engine fuel derived from waste plastics by thermal treatment," Fuel, vol. 81, no. 4, pp. 473–481, 2002.
6. J. Walendziewski, "Continuous flow cracking of waste plastics," Fuel Processing Technology, vol. 86, no. 12-13, pp. 1265–1278, 2005.
7. Y. H. Seo, K. H. Lee, and D. H. Shin, "Investigation of catalytic degradation of high-density polyethylene by hydrocarbon group type analysis," Journal of Analytical and Applied Pyrolysis, vol. 70, no. 2, pp. 383–398, 2003.
8. J. W. Park, J. H. Kim, and G. Seo, "The effect of pore shape on the catalytic performance of zeolites in the liquid-phase degradation of HDPE," Polymer Degradation and Stability, vol. 76, no. 3, pp. 495–501, 2002.
9. G. Manos, A. Garforth, and J. Dwyer, "Catalytic degradation of high-density polyethylene over different zeolitic structures," Industrial and Engineering Chemistry Research, vol. 39, no. 5, pp. 1198–1202, 2000. ·
10. A. A. Garforth, Y. H. Lin, P. N. Sharratt, and J. Dwyer, "Production of hydrocarbons by catalytic degradation of high

density polyethylene in a laboratory fluidised-bed reactor," Applied Catalysis A, vol. 169, no. 2, pp. 331–342, 1998.

11. S. Ali, A. A. Garforth, D. H. Harris, D. J. Rawlence, and Y. Uemichi, "Polymer waste recycling over "used" catalysts," Catalysis Today, vol. 75, no. 1–4, pp. 247–255, 2002.
12. J. F. Mastral, C. Berrueco, M. Gea, and J. Ceamanos, "Catalytic degradation of high density polyethylene over nanocrystalline HZSM-5 zeolite," Polymer Degradation and Stability, vol. 91, no. 12, pp. 3330–3338, 2006.
13. S. Karagöz, J. Yanik, S. Uçar, M. Saglam, and C. Song, "Catalytic and thermal degradation of high-density polyethylene in vacuum gas oil over non-acidic and acidic catalysts," Applied Catalysis A, vol. 242, no. 1, pp. 51–62, 2003.
14. K. H. Lee and D. H. Shin, "Catalytic degradation of waste polyolefinic polymers using spent FCC catalyst with various experimental variables," Korean Journal of Chemical Engineering, vol. 20, no. 1, pp. 89–92, 2003.
15. K. H. Lee, S. G. Jeon, K. H. Kim et al., "Thermal and catalytic degradation of waste high-density polyethylene (HDPE) using spent FCC catalyst," Korean Journal of Chemical Engineering, vol. 20, no. 4, pp. 693–697, 2003.
16. S. Kumar and R. K. Singh, "Recovery of hydrocarbon liquid from waste high density polyethylene by thermal pyrolysis," Brazilian Journal of Chemical Engineering, vol. 28, no. 4, pp. 659–667, 2011.
17. Petroleum Product Surveys, Motor Gasoline, Summer, Winter 1986/1987, National Institute for Petroleum and Energy Research, 1986, http://www.afdc.energy.gov/afdc/pdfs/fueltable.pdf.
18. J. Tuttle and T. V. Kuegelgen, Biodiesel Handling and Use Guidelines, National Renewable Energy Laboratory, 3rd edition, 2004.
19. A. Aboulkas, K. E. Harfi, A. E. Bouadili, M. Benchanaa, A. Mokhlisse, and A. Outzourit, "Kinetics of co-pyrolysis of

tarfaya (Morocco) oil shale with high-density polyethylene," Oil Shale, vol. 24, no. 1, pp. 15–33, 2007.

20. Y. Wu, A. V. Isarov, and C. O. Connell, "Thermal analysis of high density polethylene-maple woodflour composites," Thermochimica Acta, vol. 340-341, pp. 205–220, 1999.

Chapter 2

Evaluation of Microseismicity Related to Hydraulic Fracking Operations of Petroleum Reservoirs and Its Possible Environmental Repercussions

Abdulaziz M. Abdulaziz

Mining, Petroleum, and Metallurgical Engineering Department, Faculty of Engineering, Cairo University, Giza, Egypt

ABSTRACT

Petroleum reservoir operations such as oil and gas production, hydraulic fracturing, and water injection induce considerable stress changes that at some point result in rock failure and emanation of seismic energy. Such seismic energy could be large enough to be felt in the neighborhood of the oil fields; therefore many issues are recently raised regarding its environmental impact. In this research we analyze the magnitudes of microseismicity induced by stimulation of unconventional reservoirs at various basins in the United States and Canada that monitored the microseismicity induced by hydraulic fracturing operations. In addition, the relationship between microseismic magnitude and both depth and injection parameters is examined to delineate the possible framework that controls the system. Generally, microseismicity of typical hydraulic fracturing and injection operations is relatively similar in the majority of basins under investigation and the overall associating seismic energy is not strong enough to be the important factor to jeopardize near surface groundwater resources. Furthermore, these events are less energetic compared to the moderately active tectonic zones through the world and usually do not extend over a long period at considerably deep parts. However, the huge volume of the treatment fluids and improper casing cementing operation seem to be primary sources for contaminating near surface water resources.

INTRODUCTION

Due to the increasing demands on the traditional oil and gas resources, hydraulic fracturing became an important technology applied for enhancing production from hydrocarbon reservoirs, particularly the unconventional ones. Recently, the conjugated practices of horizontal wells with multi-stage hydraulic fracturing not only have increased the well productivity dramatically, but also lead to enormous increase in hydraulic fracturing. This increase approaches the size of massive fracture treatments carried out in the 1970 [1]. Microseismic monitoring is a new technology that typically targets the impulsive, energetic acoustic emissions to map fracture growth during hydraulic fracking stimulations. Other applications utilize these emanations to monitor the slow creeping processes within the reservoir over long period due to production operations [2]. The acoustic emissions represent the released energy during formation deformation which corresponds to small (Mw < −1.5) to medium magnitude micro-earthquakes (although high magnitudes of Mw > 0 are also reported), known as microseisms. The resulting deformation is induced by stress redistribution within the reservoir based on stress-strain interaction and may activate slippage across pre-existing structures or initiate new fractures within the stimulated reservoir volume. The analysis of the recorded microseismic data is typically useful in locating the induced fracture system [3], monitoring the geomechanical deformation [4], mapping fracture growth [5], and calculating the stimulated reservoir volume [6] by the stimulation operations.

To successfully monitor a hydraulic fracture treatment, the location of the monitoring array, determination of an adequate velocity structure, and management of noise are important issues to be considered [1]. Microseismic imaging usually employs a temporary string of eight to twelve triaxial geophones in a monitoring well located close to the treatment well to observe the creation the of induced fracture systems within its environ (Figure 1).

To capture good quality spectral responses with minimal signal interference in a microseismic record, the downhole geophones should possess sufficient sensor response, minimal tool resonances, and suitable frequency response. In addition, the acquisition system should enable sampling rate between 0.5 and 0.25 msec that corresponds to Nyquist frequency 1000 and 2000 Hz respectively. Seismic attenuation is usually encountered at high frequency signals from far events compared to the low frequency and near ones that are usually mitigated during signal processing. In most cases, microseismic monitoring utilizes downhole sensors, while near surface sensors are in some cases deployed. Being close to the source in borehole microseismic, the depth of the microseismic event is usually accurately determined, but surface arrays proved to be more efficient in determining the spatial location of hypocenter. Dynamic microseismic images are obtained as a live streaming to the fracture propagation using the time history of the microseismic activity [7]. Later, additional seismic source attributes, such as strength or magnitude, are calculated and the complete record is processed to elucidate information about the travel path, e.g. anisotropy and velocity tomograms [8].

It is well-documented that long-term injection of fluids into deep formations induces earthquakes [9] [10]. This is particularly true in regions susceptible to tectonic activities such as waste chemicals injection in the Rocky Mountains [11] , water injection in geothermal plants [10] [12] , water flooding operation for hydrocarbon production optimization [13] , and water disposal in mining industry [14] . Recently a considerable attention and arguments have been raised concerning the seismicity associating hydraulic fracturing operation [11], particularly, their role in contaminating the shallow aquifer systems that could be the main water supply for domestic use. Since such operations are not risk-free, several incidents of probable water resources contamination have been reported due to hydrocarbon production stimulation by hydraulic fracking [15] [16].

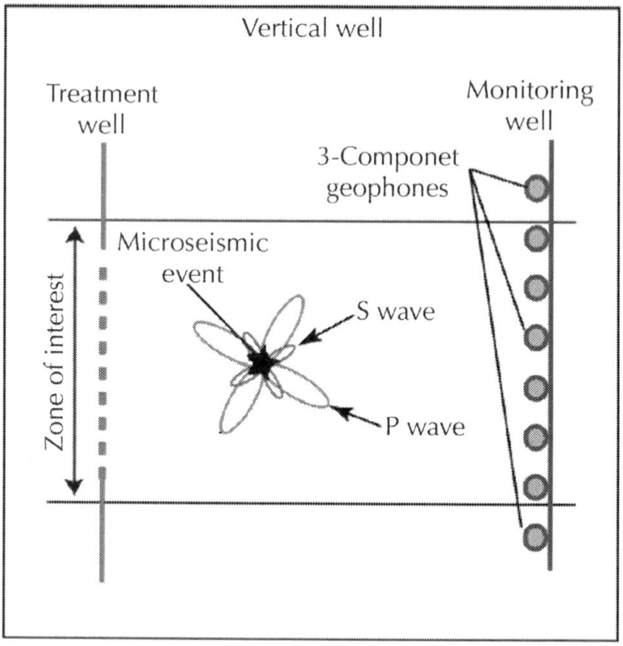

(a)

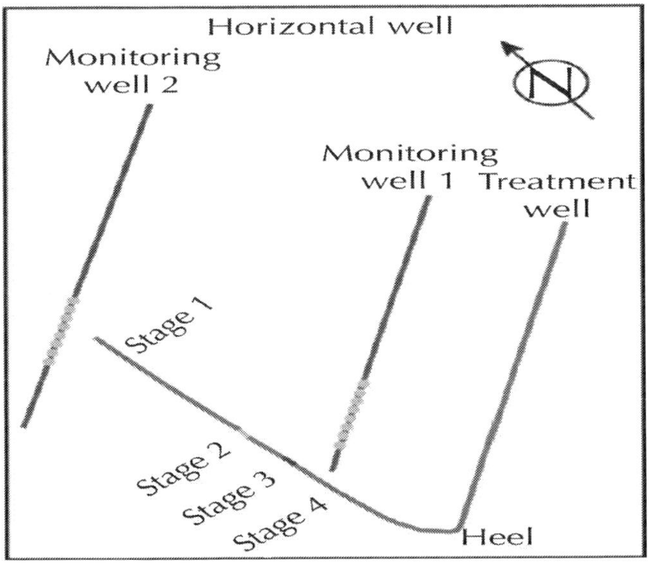

(b)

Figure 1: Typical microseismic array deployed in vertical well (a) and horizontal well (b).

Drinking water could be contaminated during hydraulic fracking operations through hidden pathways connecting the producing horizon with shallow aquifer, casing-cement failure, and flowback water used during the treatment. Reference [17] evaluated the potential impacts associating gas-well drilling and fracturing in Marcellus Shale on shallow groundwater systems in northeast Pennsylvania and upstate New York using groundwater analysis. The dissolved-methane concentrations and carbon and hydrogen isotope ratio documented the existence of thermogenic methane in numerous wells with high concentrations (a potential for explosion hazard) close to active natural-gas wells [18]. However, no signs of brine mixture from deeper formations or traces frack treatment fluids were detected. Such observations could be supportive for gas contamination from a near surface origin such as wellbore cement failure, but the deeper origin cannot be entirely excluded due to the extremely low density of methane that enables swift gas to flow across probable hidden pathways, compared to brine and frack treatment fluids. In this paper a detailed discussion on microseismicity associating hydraulic fracking is introduced to assess the potential of hydraulic fracturing to induce minute earthquakes that may jeopardize the surface and near surface environment using microseismic records of different fields, mostly from North America. The results of this research may help recognize the potential/role of hydraulic fracturing of tight petroleum reservoirs in contaminating the potable groundwater aquifers.

METHODS

Faults are physically static if the in-situ stresses are creating enough frictional forces along fault planes. Fluid injection results in shear stresses within the rock by increasing the pore pressure and therefore weakening the rock fabrics. When the shear stress increases enough to overcome the in-situ stresses, the rock initiates a fracture followed by a slip or directly starts slip on a pre-existing fault plain, resulting in an earthquake.

Reference [19], approximated the maximum static friction ($F_{max\ static}$) by:

$$F_{\text{max static}} = \mu_{\text{static}} \times F_{\text{normal}} \quad (1)$$

Where: μ_{static} is the static friction coefficient; F_{normal} is the normal force.

Since microseismic records represent a graphical demonstration to stress decay, fracture geometry and growth behavior can be identified using standard earthquake seismology principles [13] [20]. To evaluate the seismicity associating hydraulic fracturing operation, it is essential to calculate the magnitude for each recorded event and compare the calculated values with injection parameters. Seismic moment can be calculated using source parameters with several techniques, the simplest of them was introduced by reference [21]. He utilized Fourier transform of S-wave displacement to estimate the event moment and radius. In this research, moment and other source parameters data are calculated using Brune's method.

Reference [21] calculated the Moment (M**o**) and the source radius (r**o**) by

$$Mo = \frac{4\Pi \rho Vs^3 R\Omega o}{Fc} \quad (2)$$

Where: **ρ** is the density, Vs is the shear velocity, R is the distance from the receivers to the event, **Ω**o is the low-frequency amplitude of the displacement spectrum, and Fc is a radiation pattern factor. Since the hydraulic fracturing occurs within a small intervals of the producing horizon, the source-receiver distance {R} and radiation pattern factor {Fc} are most likely the influential parameters in Equation (2) to determine the Mo value.

To determine **Ω**o value for each event, the amplitude spectrum is plotted versus frequency of the microseism after correcting for attenuation. Then the corner frequency is graphically identified by the intersection of the power-law decay at high frequency with the line that approximately represents the low frequency amplitude (Figure 2). The intersection of those two lines indicates the likely corner frequency, which is approximately 350 Hz in the example shown in Figure 2, and the low-frequency amplitude of a little less than 2E−10 m-sec.

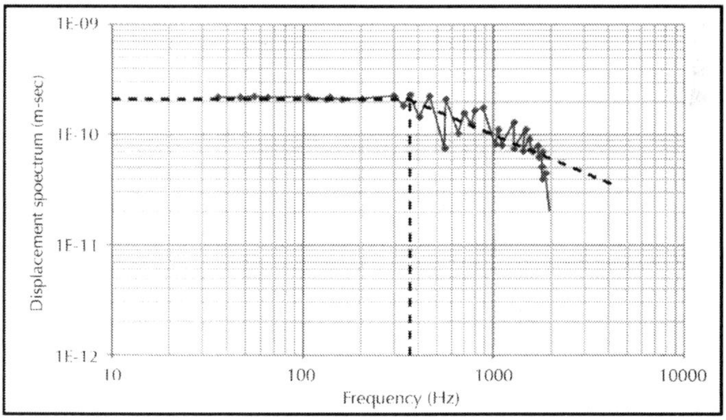

Figure 2: Graphical determination of the corner frequency using amplitude spectrum and frequency after correction for attenuation.

Reference [21] approximated the source radius (ro) as

$$ro = \frac{KcVs}{2\Pi fc} \qquad (3)$$

Where: Kc is a constant, and fc is the corner frequency. The value of the constant Kc has been used to equal (~2.2) by Reference [21] that provides, a conservative value of the size of the slippage plane.

The seismic energy (E) released during the slippage along the fault plane is approximated by Reference [22] as a function of seismic moment by:

$$E \approx Mo/20,000 \qquad (4)$$

Analogous to Richter scale of earthquakes, the moment magnitude (Mw) is a more convenient value to represent seismic moment and/or seismic energy that are obtained by:

$$Mw = \frac{2}{3}\left(\log(Mo) - 16.1\right) = \frac{2}{3}\left(\log E - 11.8\right) \qquad (5)$$

Where Mo and E in this equation are expressed in dyne-cm.

The slip displacement and area can be determined as a function of the seismic moment by:

$$Mo = \mu dA \qquad (6)$$

Where: **μ** is the shear modulus of the rock (typically 2.2×10^6 psi for shale), d is the slip distance, and A is the slippage area.

In the present work the calculated moment magnitude from different unconventional reservoirs is subjected to statistical analysis and plotted against various parameters related to the reservoir (e.g. depth) and treatment parameter (e.g. injection rate and volume). This helps understanding the capability of hydraulic fracturing operation to disturb the surface and subsurface environment at the vicinity of stimulated wells.

RESULTS AND DISCUSSION

Microseismicity and Treatment Parameters

Thousands of fracturing stimulations have been monitored using a microseismic technique, in which broad variation in the magnitudes of the recorded seismicity is documented. Based on the estimated microseismic magnitudes calculated by Brune's method [21] described earlier (Equations (2) to (7)), a plot of these values versus depth was constructed for Barnett Shale reservoir (Figure 3, data from Reference [1]). Although the data points are relatively scattered in this plot, there is a trend of increasing the number of high-magnitude events, shown as dense clusters, with increasing the depth (Figure 3). This reflects the increase of stress in the reservoir rock due to compaction with depth which produces higher strain response in the form of higher microseisms. Another important trend is the linearity of events in either cross-sectional or planar views that could be respectively attributed to numerous small displacements along bedding plains or fault plains. Generally, the majority of microseismic events showed magnitudes less than −0.8 for the shallow zones (less than 2000 m). These values slightly increase to −0.4 with increasing the depth of the target zone (Figure 3 and Figure 4). In few cases microseisms may record Mw values close to 0.5 especially in highly fractured formations such as Barnett Shale (Figure 3) with increasing the depth of the stimulated target. In addition to compaction effect, the variation in magnitudes is influenced by the prevailing geologic structures and/or density of natural fractures that generally controls stress distribution in the stimulated area. Other factors that may influence the variation in magnitude include the treatment parameters such as injection rate, fluid properties, target depth and thickness, completion characteristics, and well design [23].

The plot of number of events versus magnitude for seismic data typically follows a power law distribution described by the

Gutenberg-Richter relationship, (logN = a − bM). However this plot for both Barnett Shale and Cotton Valley reservoirs data set (Figure 4 and Figure 5) had normal distribution patterns for different monitored depth intervals. This could be attributed to inadequate capacity of the recording system to detect smaller magnitudes, Mw < −1.4, at large distance from the monitoring well due to higher signal to noise ratio. In addition, the high number of events cluster towards the right-hand side (high Mw values, Figure 5upper part) but the high numbers of recorded events are slightly shifted to the left-hand side of the plot in deeper stimulated intervals (low Mw values, Figure 5 lower part). This shift indicates different magnitudes of completeness (Mc) with deeper monitoring experiments achieving better Mc. Considering a high magnitude event to represent the foot print of initiation of a main fracture or at least activation of some pre-existing ones, the high numbers of strong microseismic events in shallow zones compared to smaller magnitudes reflect the initiation of a major fracture. However, the main fractures in deeper parts propagate for larger distance into the formation since the event magnitudes of the shallow zone are relatively lower than the deep zone and low magnitude (Mw < −1.4) events are distinctly recorded in deeper stimulated intervals (Figure 5).

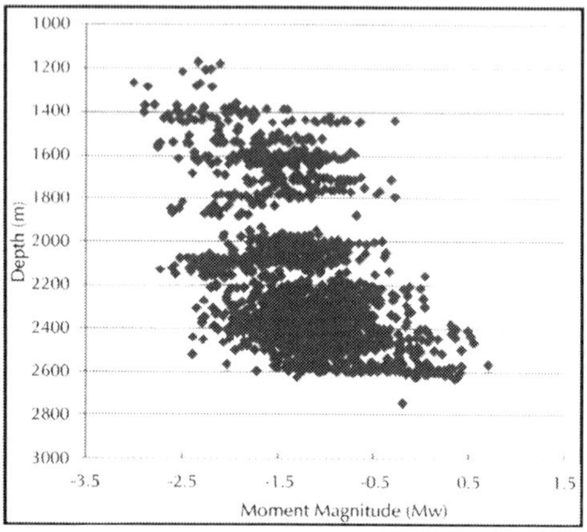

Figure 3: microseismic magnitudes calculated by Brune's method plotted versus depth in Barnett shale gas reservoir, Data from Reference [1].

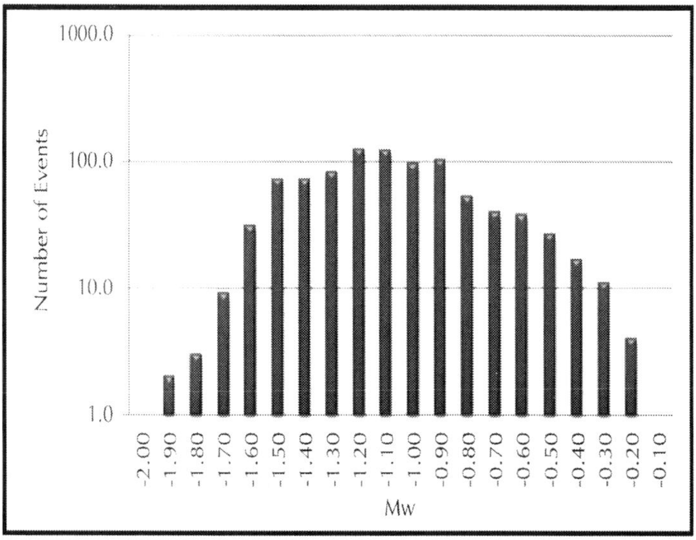

Figure 4: 994 microseismic events recorded by two monitoring well at depth between 2756 and 2838m in Cotton valley sandstone gas field, East Texas Data from Reference [24].

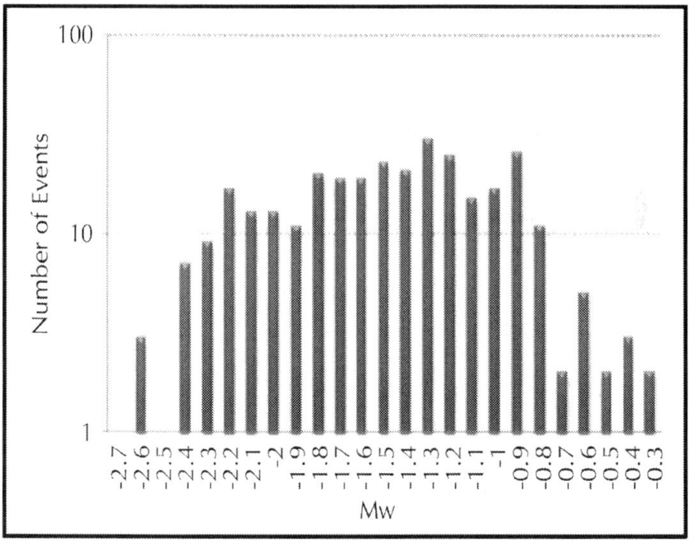

(a)

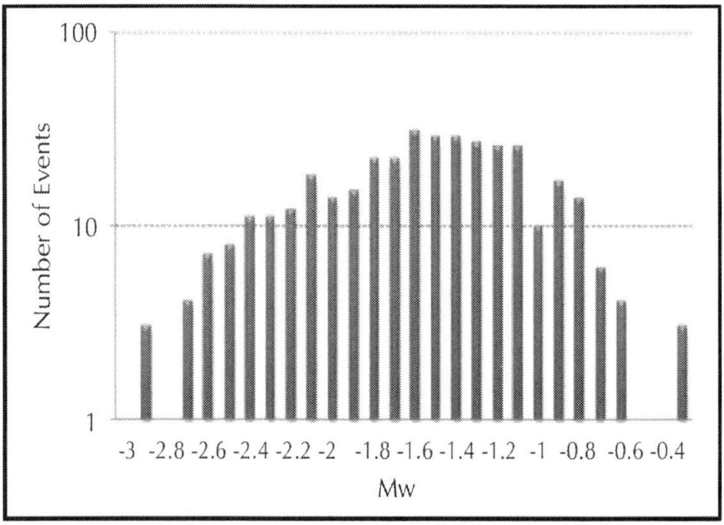

(b)

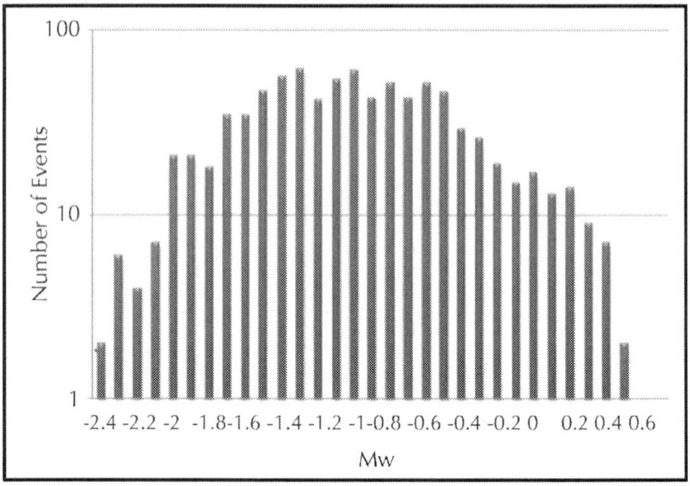

(c)

Figure 5: Plot of the number of events versus magnitudes recorded in depth intervals 1170 - 1975 m 1975 - 2200 m and 2200 - 2750 m showed in upper, middle, and lower respectively for the data shown in Figure 3 of Barnett Shale (Data from Reference [1]).

The injection rate and volume are expected to be influential factors to microseismicity, but could be the most argumentative issues of hydraulic fracking stimulations especially, handling and treatment of the flowback water. To distinguish the relationship between seismicity and both injection rate and injection volume associating hydraulic fracturing stimulations, comprehensive data sets from Barnet shale [1] have been digitized for statistical analysis and interpretation. Figure 6 shows two plots of these data; the first set shows injection rate (m^3/minute) versus the recorded moment magnitude and the second displays average magnitude versus the number of the recorded events at the corresponding rate. Although there is a notable increase in the number of recorded events and the average recorded magnitude with increasing the injection rate at the early stages of the treatment (up to 8 m^3/minute), such a trend is totally inversed with dramatic changes in rate for the number of events compared to the average recorded magnitude (Figure 6).

The relatively large average magnitudes, reported at injection rate above 10 m^3/minute, are attributed to the small number of low magnitude events and the abundant moderate magnitude values. The maximum number of events was recorded at injection rate of 6.0 m^3/minute, while the higher average value came slightly later at approximately 8 m^3/minute. These observations are common in highly fractured tight formations such as those existing in East Texas, where very small displacements causing small magnitudes are numerous near the treatment well vicinity and remarkably decrease as moving away [25]. Such events are replaced by moderate magnitudes of much lower numbers which are usually arranged parallel to the general structural trends. This may indicate the development of larger stimulated reservoir volume near the treatment well replaced by convey channels away. Furthermore, the relatively larger microseismic magnitudes ($Mw > 0$) are obviously restricted to the same interval of injection rate which seem critical to this geologic media to create fracture systems. These systems act as effective conduits enough to mitigate the influence of additional increase in injection rate [26]. This may explain the large increase in production rates associating minor seismicity at the first few weeks after treatment (draining the stimulated reservoir volume near the well bore) followed by oscillating pattern that finally collapse to the normal low production rates as clearly shown in Figure 7. Alternatively, the plot of injection volume versus both number of events and average magnitude values (Figure 8) recorded in two shale reservoirs, Marcellus and Barnett, showed marked increase in seismicity during the first 1600 - 2300 m^3 of injected fluids. Through this range, the maximum number of events is recorded after injecting ~1300 m^3 of the treatment fluids. Comparing the two plots of Figure 8, the increasing number of events indicates a clear response to the stimulation process at the early stage in which significant displacements along pre-existing fractures and initiation of new fracture systems are expected. After that, the number of events is continually decreasing as a result of the contemporaneous improvement of the local permeability by the fracking process. Such response enhances the local permeability that requires additional volume of treatment fluids and adjustment to injection rate to be able to start new significant seismicity. Despite the clear response of injection volume to seismicity, the average recorded magnitudes were less obvious particularly for Marcellus Shale that shows minute changes through the record (Figure 8). The high average values reported at the later stages is attributed to the significant high proportion of medium-value events compared to the low-value events. It is common for most basins to show similar pattern to that described in Figure 8 with various rates of changes. These results disagree with the conclusion of [1], who reported the lack of evidence for either rate or volume effects on the size of the microseisms.

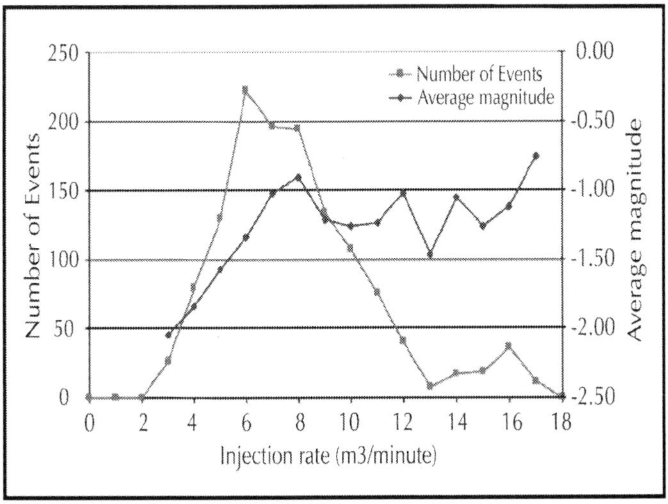

Figure 6: Plot of injection rate (m^3/minute) versus Moment magnitude recorded during Barnett shale gas stimulation from different fields in East Texas (Data from Reference [1]).

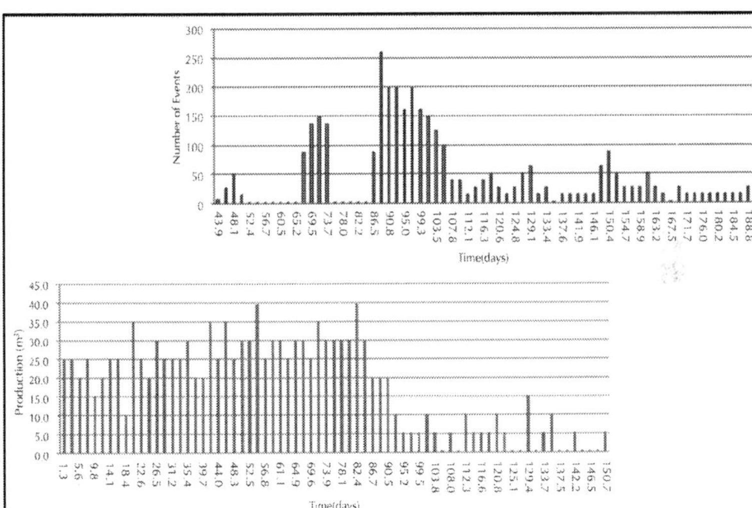

Figure 7: Comparison between production versus time plot and number of events versus time plot for a well in Clinton County, Kentucky Data from Reference [27].

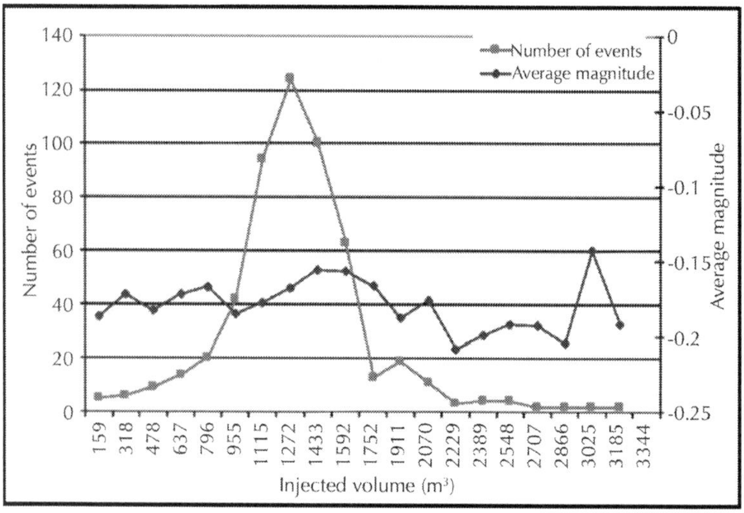

(a)

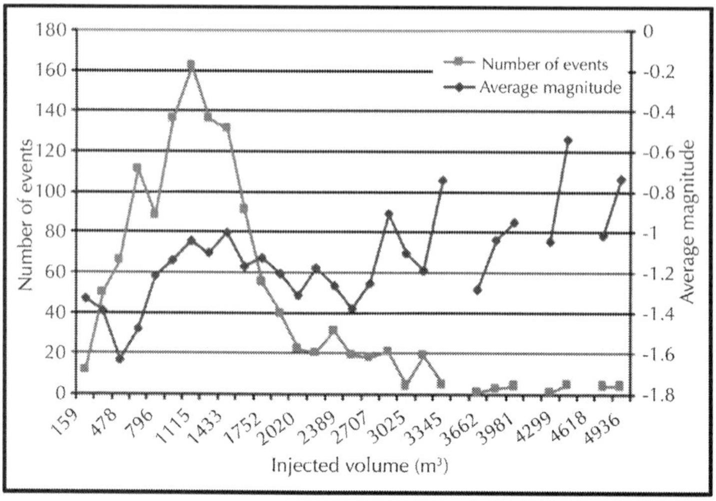

(b)

Figure 8: Plot of the injected volume of the treatment fluids (m^3) versus both number of events and average magnitude recorded during hydraulic fracturing stimulation in Marcellus (upper) and Barnett (lower) shale Data from Reference [1].

It is a common practice in hydraulic fracturing operations to begin with a minifrac test and adjust the injection parameters to fit the reservoir condition. Microseismicity usually starts shortly (an hour or so) after the commence of the main treatment (Figure 9). However aseismic events, too small magnitudes to be recorded by downhole tools, most likely exist. On comparing the number of microseismic events to the contemporaneous well head pressure (Figure 9), it is obvious that higher microseismicity is encountered after pressure buildup (between hour 10 and 11), which could have started a little earlier. But as a general pattern, pressure build up with numerous microseismicity is only observed at the early and middle stages of the main treatment (Figure 9). In addition, microseismicity are recorded for few hours after the cease of injection while only few events are encountered during the following weeks of the treatment. In a six months monitoring program by Reference [27] to evaluate the seismicity associating oil production from a tight carbonate reservoir (2% porosity), post a successful hydraulic fracturing job, a total 3200 events of average Mw = −1.2 were recorded during the production of a total of 1250 m^3 of oil (Figure 7). The recorded microseismicity during this monitoring program was practically restricted to the stimulated reservoir horizon which indicates a direct relation to recovered-oil production.

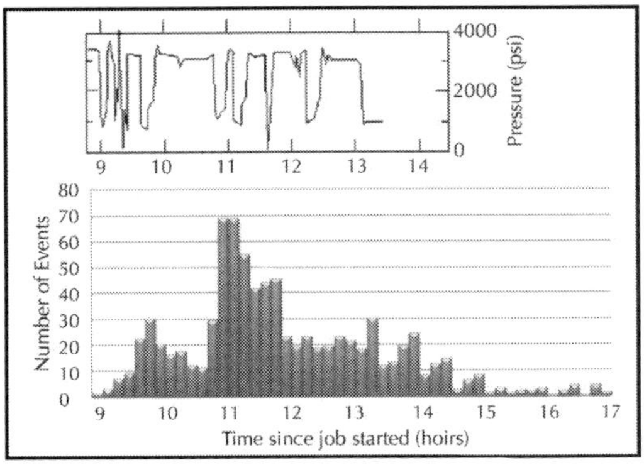

Figure 9: Plot of number of events and Pressure (psi) versus time (hours) since injection started in well CPU2-2, Giddings, Texas Data from Reference [27].

A typical seismicity pattern following hydraulic fracturing (Figure 7) shows insignificant microseismic activity in the first 3 months of putting the well on production with a remarkable increase in production rate that usually fluctuates within 30 to 50 % of the total increase in well productivity. By the end of the high production records (90 days period, Figure 7), an unusual increase in seismicity was recorded reflecting abnormal changes taking place within the reservoir followed by continuous decrease in oil production (Figure 7). The recorded seismicity associating diminished well production represents a massive fracture closure or possible fault activation. This may be due to fluid withdrawal from the stimulated reservoir volume that was supporting the overburden pressure and eventually reached failure along the weak zones. There is few recorded seismicity followed these significant modification within the reservoir architecture that extend between days 107 and 170 (Figure 7). This period corresponds to a transition stage towards the stable reservoir condition in which well productivity is dramatically reduced. This pattern represents a typical tight reservoir behavior following hydraulic fracturing stimulations. Such observations indicate a direct relation between the induced seismicity and the efficiency of hydraulic fracking stimulation operations.

The reservoir stimulation efficiency as well as the resulting seismicity varies dramatically among reservoirs and within the same reservoir. However, post-frack high productivity correlates generally with seismicity over wide zones within the producing horizon. Figure 10 shows the cumulative production of two wells that was stimulated by hydraulic fracturing within the same reservoir. In this case the high productivity well showed seismicity over an area 4 - 5 times the stimulated area of the second well. This change in the stimulated reservoir volume is attributed to the reservoir rock competency and durability. Competent rocks favor the development of a single fracture over long distance but durable rocks (highly fractured), prefer fracture propagation over large area rather than long distance [23] [27] . The energy infused into the formation could be dispensed either as numerous micro-fractures or displacement along a fault plane with distinct microseismic record describing each case. Typically, microseismic activity and strength diminishes at the end of the injection treatment and the seismic moment curve stabilizes at approximately minimal values. However, in experiments incorporating fault activation, seismic deformation begins to diminish after the end of injection followed by sharp increase in seismicity that may extend for several tens of minutes before it completely stabilizes. Reference [20] characterized fault activation during hydraulic fracturing of a gas well penetrating a thrust fault. Two different deformation mechanisms: during (fracturing) and after (fault activation) injection were distinctly recognized with two different Frequency-magnitude characteristics. In addition, the P/S amplitude ratio for post injection seismicity are relatively high compared to those recorded during the pumping, suggesting deformation at different orientation that probably include activation of a nearby fault [20] .

Environmental Repercussions of Hydraulic Fracking

To be felt on land surface, the earthquake's magnitude should be roughly +3 with approximate energy 2.033×10^9 N-m [1]. This magnitude is too large compared to the magnitude microseism with Mw = −2 and energy 68 N-m. In addition, the detailed discussion of the relation of microseismicity to the depth (Figure 3), magnitude (Figure 4 and Figure 5), injection rate and volume (Figure 6 and Figure 8), and well production (Figure 7 and Figure 10) indicated a conservative framework and limited magnitude of energy. This indicates that the microseismicity associating hydraulic fracking are only located within and/or close to producing horizons that are separated by thousands of rock masses from shallow aquifers. Even microseismicity associating fault activation, within the proximity of a treatment well, may hit nearby deep aquifer [8]. This seems hard to jeopardize near surface groundwater resources unless such faults are naturally good conduits (in such cases fracking only increases contamination that is originally existed). Accordingly, all recorded events do not directly jeopardize the surface environment or the ecology as they cannot be detected except by highly sensitive instruments. Furthermore, hydraulic fracking operation is not expected to induce destructive earthquakes. The USGS inventory of earthquake data since 1900 did not show significant changes compared to the pre-existing data. Therefore, the mechanical hydraulic fracking seems unsuspicious for contaminating the drinking water resources. However, there are substantial hazards entailing the huge volume of treatment fluids that return to the surface charged with chemicals, radioactive materials, and salts. Such fluids are easy to reach nearby rivers and shallow aquifers [28]. In addition, the other possible pathway for gas transport to surface aquifer may take place through the vertical wellbore with poor cement job especially the surface part and/or across the abandoned old well with post fracking cross communication with the producing wells. This indicates the importance and the rising need for better management of shale-gas extraction through integration of the available data, particularly microseismic technology, to improve the industry practices in drilling and hydraulic fracking.

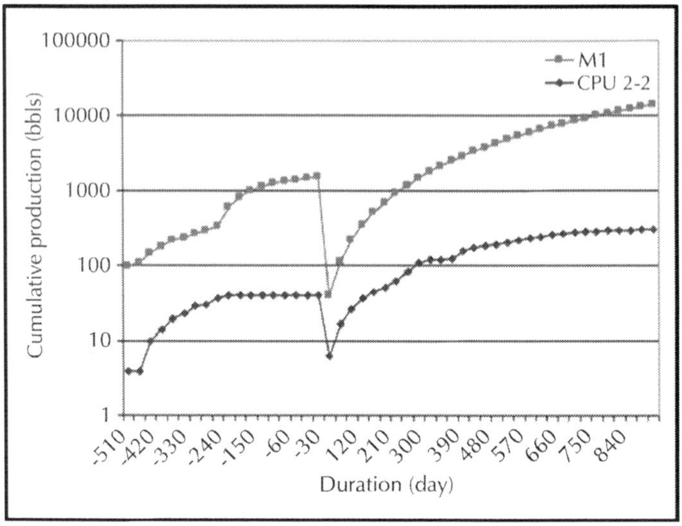

Figure 10: Cumulative production Preand Post-stimulation of two oil wells in Giddings field, Texas Data from Reference [27].

CONCLUSIONS

The seismicity recorded during hydraulic fracking operations is strongly dependent on the reservoir geology and the parameters of hydraulic treatment. The present study showed that the maximum number of recorded events was recorded at injection rate of 6.0 m^3/minute, while the higher average value came slightly later at approximately 8 m^3/minute and a marked increase in seismicity is recorded during the first 1600 - 2300 m^3 of injected fluids. In addition, the average microseism magnitude ($Mw = -2$) and energy (68 N-m) presented in this study may indicate limited role of hydraulic fracking stimulations in initiating any destructive earthquakes. This indicates that mechanical processes of hydraulic fracking operations do not directly jeopardize the surface environment or the ecology, but subsurface hazards cannot be totally excluded. The huge volume of treatment fluids that return to the surface charged with chemicals and improper well completion, particularly wellbore cement, represent the important hazardous source. The results of the present study helps understand the behavior of hydrocarbon reservoirs to hydraulic fracking and production operations that may enable a better reservoir management and improve the industry practices in drilling and hydraulic fracking.

ACKNOWLEDGEMENTS

The author acknowledges very much the authors contributed with the data used in this research and the efforts of their microseismic crews. Also the fruitful suggestions and careful revisions made by Professor Dr. Abdel-Zaher Abouzeid (Professor of Mining Engineering) and Professor Dr. Fouad Khalaf (Professor Petroleum Engineering) at Faculty of Engineering, Cairo University are greatly appreciated.

REFERENCES

1. Warpinski, N.R., Du, J. and Zimmer, U. (2012) Measurements of Hydraulic-Fracture-Induced Seismicity in Gas Shales. SPE Hydraulic Fracturing Technology Conference, Woodlands, 6-8 February 2012, Article ID: 151597.
2. Maxwell, S.C., Urbancic, T., Steinsberger, N. and Zinno, R. (2002) Microseismic Imaging of Fracture Complexity in the Barnett Shale. Proceedings, Society of Petroleum Engineers Annual Technical Conference, SPE Annual Technical Conference and Exhibition, San Antonio, 29 September-2 October 2002, Paper # 77440.http://dx.doi.org/10.2118/77440-MS
3. Mohammad, N.A. and Miskimins, J.L. (2012) A Comparison of Hydraulic-Fracture Modeling With Downhole and Surface Microseismic Data in a Stacked Fluvial Pay System. SPE Production & Operations Journal, 27, 253-264. http://dx.doi.org/10.2118/134490-PA
4. Maxwell, S.C, Du, J. and Shemeta, J. (2008) Monitoring Steam Injection Deformation Using Microseismicity and Tiltmeters. The 42nd U.S. Rock Mechanics Symposium (USRMS), San Francisco, June 29-July 2, 2008, 8.
5. Warpinski, N.R., Mayerhofer, M.J., Vincent, M.C., Ceramics, C., Cipolla, C.L. and Lolon, E.P. (2009) Stimulating Unconventional Reservoirs: Maximizing Network Growth While Optimizing Fracture Conductivity. Journal of Canadian Petroleum Technology, 48, 39-51.http://dx.doi.org/10.2118/114173-PA
6. Mayerhofer, M.J., Warpinski, N.R., Cipolla, C.L., Walser, D. and Rightmire, C.M. (2010) What Is Stimulated Reservoir Volume? SPE Production & Operations, 25, 89-99.http://dx.doi.org/10.2118/119890-PA
7. Ajayi, B., Walker, K., Sink, J., Wutherich, K. and Downie, R. (2011) Using Microseismic Monitoring as a Real Time Completions Diagnostic Tool in Unconventional Reservoirs:

Field Case Studies. SPE Eastern Regional Meeting, Columbus, 17-19 August 2011, 10.

8. Le Calvez, J.H., Klem, R.C., Bennett, L., Erwemi, A., Craven, M. and Palacio, J.C. (2007) Real-Time Microseismic Monitoring of Hydraulic Fracture Treatment: A Tool To Improve Completion and Reservoir Management. SPE Hydraulic Fracturing Technology Conference, College Station, 29-31 January 2007, 7. http://dx.doi.org/10.2118/106159-MS
9. Segall, P. (1989) Earthquakes Triggered by Fluid Extraction. Geology, 17, 942-946.http://dx.doi.org/10.1130/0091-7613(1989)017<0942:ETBFE>2.3.CO;2
10. Smith, J.L.B., Beall, J.J. and Stark, M.A. (2000) Induced Seismicity in the SE Geysers Field. Geothermal Resources Council—Transactions, 24, 24-27.
11. Nicholson, C. and Wesson, R.L. (1990) Earthquake Hazard Associated with Deep Well Injection—A Report to the U.S. Environmental Protection Agency. U.S. Geological Survey Bulletin 1951. Denver.
12. Fehler, M. (1989) Stress Control of Seismicity Patterns Observed during Hydraulic Fracturing Experiments at the Fenton Hill Hot Dry Rock Geothermal Energy Site, New Mexico. International Journal of Rock Mechanics and Mining Sciences & Geomechanics, 26, 211-219. http://dx.doi.org/10.1016/0148-9062(89)91971-2
13. Zoback, M.D. and Harjes, H.P. (1997) Injection-Induced Earthquakes and Crustal Stress at 9 km Depth at the KTB Deep Drilling Site, Germany. Journal of Geophysical Research, 102, 18477-18491. http://dx.doi.org/10.1029/96JB02814
14. Ake, J., Mahrer, K., O'Connell, D.O. and Block, L. (2005) Deep-Injection and Closely Monitored Induced Seismicity at Paradox Valley, Colorado. Bulletin of the Seismological Society of America, 95, 664-683. http://dx.doi.org/10.1785/0120040072
15. Fakete, J. and R. Penty (2011) Environment Canada to Study Hydraulic Fracturing. Post Media News and Calgary

Herald. http://www.ernstversusencana.ca/wp-content/uploads/2011/08/2011-09-21-Ottawa-enters-debate-over-shale-gas-development-Environment-Canada-to-study-hydraulic-fracturing.doc

16. Chafin, T.D. (1994) Source and Migration Pathways of Natural Gas in Near-Surface Ground Water beneath the Animas River Valley, Colorado and New Mexico. USGS Water Resources Investigations Report 94-4006. http://pubs.er.usgs.gov/usgspubs/wri/wri944006
17. S.G. Osborn, Vengosh, A., Warner, N.R. and Jackson, R.B. (2011) Methane Contamination of Drinking Water Accompanying Gas-Well Drilling and Hydraulic Fracturing. Proceedings of the National Academy of Sciences of the United States of America, 108, 8172-8176.http://dx.doi.org/10.1073/pnas.1100682108
18. Revesz, K.M., Breen, K.J., Baldassare, A.J. and Burruss, R.C. (2010) Carbon and Hydrogen Isotopic Evidence for the Origin of Combustible Gases in Water Supply Wells in North-Central Pennsylvania. Applied Geochemistry, 25, 1845- 1859.http://dx.doi.org/10.1016/j.apgeochem.2010.09.011
19. Tipler, P.A. (1995) Physics for Scientists and Engineers. 3rd Edition, Worth Publishers.
20. Maxwell, S.C., Waltman, C.K., Warpinski, N.R., Mayerhofer, M.J. and Boroumand, N. (2009) Imaging Seismic Deformation Induced by Hydraulic Fracture Complexity. SPE Reservoir Evaluation & Engineering Journal, 12, 48-52.http://dx.doi.org/10.2118/102801-PA
21. Brune, J.N. (1970) Tectonic Stress and the Spectra of Seismic Shear Waves from Earthquakes. Journal of Geophysical Research, 75, 4997-5009.http://dx.doi.org/10.1029/JB075i026p04997
22. Kanamori, H. (1977) The Energy Release in Great Earthquakes. Journal of Geophysical Research, 82, 2981-2987. http://dx.doi.org/10.1029/JB082i020p02981

23. King, G.E. (2010) Thirty Years of Gas Shale Fracturing: What Have We Learned? SPE Annual Technical Conference and Exhibition held in Florence, Italy, 19-22 September 2010, 50.
24. Urbancic, T.I., Shumila, V., Rudledge, J.T. and Zinno, R.J. (1999) Determining Hydraulic Fracture Behavior Using Microseismicity. The 37th U.S. Symposium on Rock Mechanics (USRMS), Vail, 7-9 June 1999, 8.
25. Abdulaziz, M. (2013) Microseismic Imaging of Hydraulically Induced-Fractures in Gas Reservoirs: A Case Study in Barnett Shale Gas Reservoir, Texas, USA. OJG, 3, 361-369.
26. King, G.E. (2012) Hydraulic Fracturing 101: What Every Representative, Environmentalist, Regulator, Reporter, Investor, University Researcher, Neighbor and Engineer Should Know About Estimating Frac Risk and Improving Frac Performance in Unconventional Gas and Oil Wells. SPE Hydraulic Fracturing Technology Conference, Woodlands, 6-8 February 2012, 80.
27. Phillips, W.S., Rutledge, J.T., Fairbanks, T.D., Gardner, T.L., Miller, M.E. and Schuessler, B.K. (1998) Reservoir Fracture Mapping using Microearthquakes: Austin Chalk, Giddings Field, TX and 76 Field, Clinton Co., KY. SPE Reservoir Evaluation & Engineering Journal, 1, 114-121. http://dx.doi.org/10.2118/36651-PA
28. Mooney, C. (2011) theTruth about Fracking. Scientific American Magazine, November 2011. http://www.scientificamerican.com/article.cfm?id=the-truth-about-fracking

Chapter 3

Environmental Consequences of Long-Term Development of Petroleum Fields, Absheron p-la, Azerbaijan, Case History

Akper A. Feyzullayev and Vagif B. Ibragimov

Geology Institute of Azerbaijan National Academy of Sciences, Baku, Azerbaijan

ABSTRACT

In this paper the overview of a level of study of the deformation processes on long-term petroleum field development and ecological consequences accompanying them is given. In more details these processes are considered for the oldest Absheron petroleum bearing region (Absheron p-la, Azerbaijan) where fields are strongly depleted and the pressure drop reaches 80% from initial values. Ecological consequences of this phenomenon are the following: development in area of petroleum fields Balakhany-Sabunchi-Ramany, Surakhany, Garachukhur, Bibi-Eybat intensive pro- cess of ground subsiding (up to 47 mm/year) and flooding, frequent incidences of curved boreholes, breaks in oil, gas, and water pipelines, and sudden kicks of water and sand, occurrence of the induced seismicity (Surakhany earthquake in 1937 with magnitude 6). With purpose of forecasting of geodynamic processes creation on petroleum production complexes of system of the environmental control is recommended.

INTRODUCTION

As is known large amounts of oil, gas and associate water are extracted from the subsoil when carrying out long term hydrocarbon field development. This results in the significant pressure decline in the reservoirs which, according to different estimates, is 50% - 80% below hydrostatic pressure [1]. Such pressures are classified as abnormally low reservoir pressures (ALRP) [2].

The reservoir pressure decline leads to the increase of the effective pressure which in its turn causes deformation processes such as decrease of rock porosity and permeability as well as additional compaction [3]. This process finds its reflection in the change of the environmental situation in the oil fields which is manifested in the soil subsidence [4] [5], under flooding [6] , appearance of "induced" seismicity [7] ; that finally can cause damages for the infrastructure (well, platform and pipeline failure and others) and require large investments for its rehabilitation [8] . In this paper an overview of the problem study in the world and analysis of some of its aspects for the Absheron peninsula (Azerbaijan) (Figure 1), where oil and gas fields have been developed for more than a century and are currently considerably depleted, is provided.

OVERVIEW OF PROBLEM

Land Subsidence and Flooding

There are distinguished three types of manmade extraction of subsoil fluids which under favorable geological conditions can cause significant LS: 1) oil, gas and associated water production; 2) hot water or steam extraction for the purpose of using geothermal energy 3) ground water extraction. Each of them causes maximal LS nearly of the same order of magnitude. Thus, for instance, the LS at the well-known Wilmington oil field (California State, USA) reached 9 meters. Hot water extraction for the purpose of using geothermal energy in Wairakei (New Zealand) led to 6 - 7 meters of subsidence, and ground water extraction in Mexico (Mexico) and in San Joaquin valley in California (USA) caused the 9 meter LS [9] .

Land subsidence connected with intensive hydrocarbons production is observed in many regions of the world. Such phenomenon was for the first time recorded in 1926 at the Goose Creek oil field not far from Houston (Texas State, USA). Here the oil production from loose sandstones and silts from the depth of 350 - 1400 m (thickness of the oil bearing interval more than 300 m) during the period of 1917-1925 led to 1 m subsidence [10] - [12] . Later, the similar effect was identified at other areas. Particularly, the rate of changeable in space and time LS at Belridge and Lost Hills oil fields in California (USA) reaches 30 - 40 cm annually [13] . In Houston- Galveston (Texas area, USA), where there are at least 110 oil and gas fields, 29 fields have been crossed by the lines of relevelling. Six of them have been discovered to have local land subsidence: Alco-Mag, Chocolate Bayou, Goose Creek, Hastings, Mykawa and South Houston, whereas the LS rates (up to 120 mm/ year) exceeded the natural subsidence rates estimated to be 13 mm/ year [12] . It is thought that such rapid LS rates have been caused by the large scale removal of fluids forming a large subsidence cup [14].

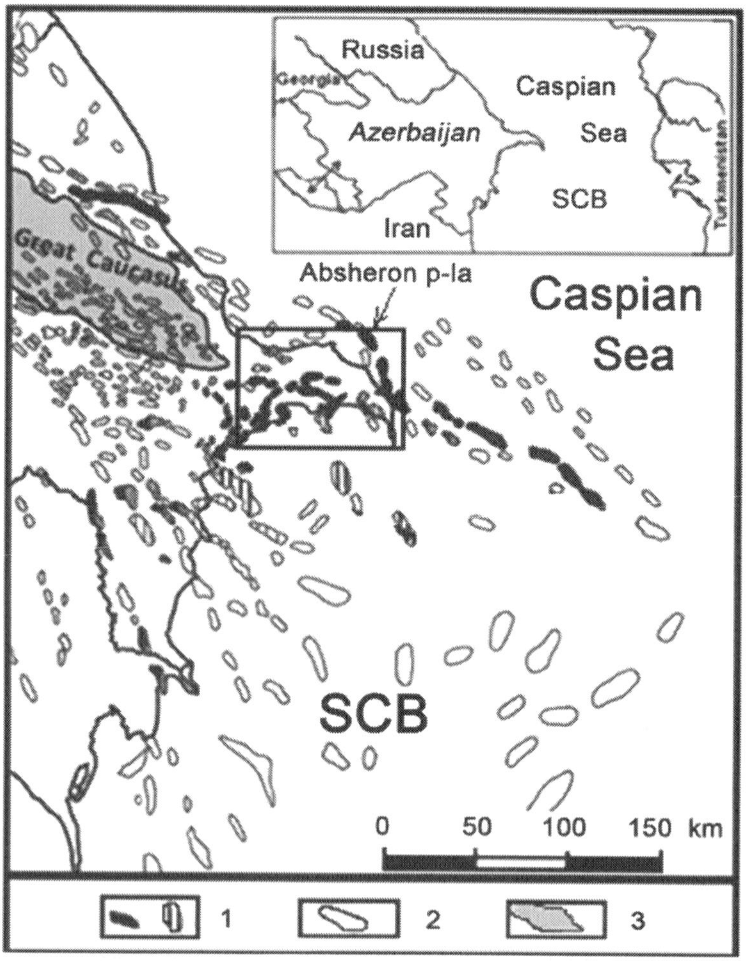

Figure 1: Location of Absheron p-la, Azerbaijan. 1: petroleum fields; 2: structures; 3: mountains.

This phenomenon is thought to have two main reasons: a) natural, related with geological process of sediments compaction, and b) manmade ones. The first ones though widely spread are characterized by relatively low rate of LS compared with the second ones. For instance, the study of the history of development and reservoir pressure at some large oil and gas fields in Louisiana state (USA) suggested that the rates of the man-in- duced LS reach 9 mm/

year or 23 mm/year (depending on whether these data have been obtained based on the results of releveling or sea level records), whereas geological rates of LS come up to about 3 mm/year [4] [15] . The rates of LS at the Port Neches field in Texas (USA) in the period of maximal petroleum production and discovered that the computed rate of land subsidence (30 mm/year) are 3 order higher than the geological rates of LS at the coastal valley [15] . The average historical rate of LS in the Mississippi delta is 12mm/year [16] which is higher than the average geological rates of LS [17].

Considerable LS at the developed hydrocarbon pools is observed under the presence of all or some of the following conditions: 1) essential drop of the reservoir pressure during the hydrocarbons production [22]; 2) hydrocarbons production from wide depth interval [18] ; 3) oil or gas (separately or together) are contained in slightly consolidated or slightly cemented rocks; 4) reservoirs have a relatively shallow depth occurrence and considerable extension.

Collection, classification and analysis of the published data on more than 130 hydrocarbons pools under production have discovered the LS occurrence from tens of centimeters to several meters [19] [20]. Special studies have established the following characteristic features of the considered phenomenon:

-The rapid decline of the hydrocarbons production leads to the reduction of LS rates. So, as a rule, the initial intensive period of the field development is characterized by higher LS rates which later slow down whereas in some cases subsidence stops.

-LS on the developed oil and gas fields has usually local character. However, hydrocarbon production from one bed on several relatively near fields can lead to the regional pressure loss and regional LS [21].

-LS can be restrained by means of reservoir pressure recovery/ maintenance. At the same time, the process of recovery can be successful if only all fault blocks work as a whole. According to some researchers actions aimed at maintaining the reservoir pressure by means of water injection into the bed and other methods applied at many fields worldwide allow to recovery pressure not exceeding

10% of initial value. It is connected with the fact that inelastic deformations of rocks accompanying the reservoir pressure decline are more significant than the elastic ones which leads to irreversible process of reservoir rocks consolidation [3]. Therefore restore the reservoir pressure to the initial value will not lead to the recovery of the initial reservoir capacity [22].

Widely spread processes of reservoir compaction and subsidence belong to the categories of dangerous natural phenomena which are inevitably accompanied by serious environmental and economic consequences. The economic consequences include flooded territories due to LS especially in coastal areas where even slight LS can cause flooding. Such phenomenon was for the first time described for the oil field near Houston (Texas State, USA). The land subsidence which reached almost 9 meters due to the oil extraction from the main Wilmington field [9] led to the flooding of streets and piers in the city and in the Long Beach port (California, USA), caused damages to the bridges, railways and other port facilities [23] . The smaller LS (~1 m) at the Goose Creek field (Texas, USA) was enough to change the landscape from the green highland to open water. LS in Venezuela led to the huge flood at the coastal area of the Maracaibo Lake, and in Russia large oil bearing areas in Western Siberia turned into swamps due to similar reasons [6] .

Economic consequences of LS as the result of hydrocarbons production often cause serious damages to the existing infrastructure (well, pipeline and platform failure etc.). According to one of the present day estimates the annual damage caused by LS due to hydrocarbons production within the USA comes up to more than 100 million dollars [8] . Here should also be added the so called "postponed" damage which will be quite appreciable by future generations: it has been established that 2 - 3 m of LS lead to the reduction of crop yield by 10%, 5 - 6 m—by 50%, and more than 8 m—cropland is totally destroyed [24] .

Induced Seismicity

It has been established that man's activity can also provoke emergence or intensification of seismic activity in the areas of hydrocarbons production. Such seismicity was named as man caused. In literature it is often referred to as "caused", "induced", "excited", "generated" etc. With the purpose of unification of terminology Russian scientists V. V. Adushkin and S. B. Turuntaev [7] offers all seismic events which occur due to various reasons after or in the process of man caused impact to define as induced seismicity.

The occurrence and intensity of induced seismicity at developed oil and gas fields depend on the geological-tectonic properties of the reservoir and surrounding massif, its mode of deformation, degree of blockiness and presence of heterogeneity, rates and volumes of hydrocarbons extraction and water injection [25] . According to the analysis of 200 fields situated in different regions of the world the likelihood of induced subsidence occurrence increases with the increase of depth and thickness of the developed bed as well as with the decline of reservoir porosity and permeability. Therewith, focuses of induced seismicity occur at various depths (depending on the geological structure of the massif) over and below the productive zones, and its strengthening happens in case of misbalance between the volume of extracted oil and injected liquid.

Researches show, that the interval of time between the beginning of development of petroleum fields and the provoked activization of seismic activity for gas fields is less (2 - 16 years), than for oil (7 - 39 years) [26] . The examples of induced seismicity are well known: it are disastrous earthquakes in the area of the Gazli gas field in 1976 and in 1984 with magnitudes from 6.8 to 7.3, and the Neftegorsk earthquake in 1995 with the magnitude of 7.2 - 7.6 which was the consequence of the active oil production on the Sakhalin island [27] .

RESULTS AND DISCUSSION

As is known the history of oil production on the territory of the Absheron peninsula (Azerbaijan) goes back to the middle of the XIX century when oil was extracted from shallow wells dug by hand. Later with the introduction of well drilling technology the intensity of hydrocarbons extraction from reservoirs was continuously increasing and by the end of the XX century exceeded one billion tons of oil (plus the extraction of large amounts of associated gas and reservoir water). This resulted in the abrupt decline of reservoir pressures.

The analysis suggests that in operated sites of the productive stratum of the Absheron peninsula the initial reservoir pressures which approximately corresponded to hydrostatic ones dropped from 52.3% (Gala field) to 80.6% (Chakhnaglar-Sulutepe field). Whereas the level of primary pressure drop in the reservoirs of the upper formations of the productive stratum is higher (in average about 66%) than in the reservoirs of the lower formations (58.9%) of the productive stratum. The level of present day pressures in comparison with the primary ones is shown on the diagram of their change with depth built according to the data of about 15 fields of Absheron peninsula (Figure 2).

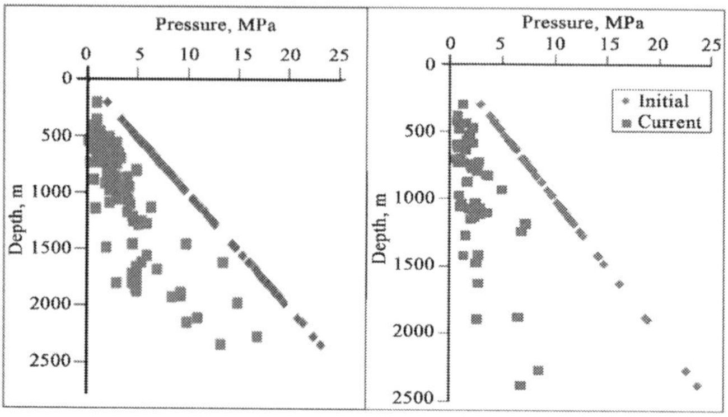

Figure 2: Change with depth of initial and current (as of 01.01.2009) reservoir pressures at the fields of Absheron peninsula (on left) and at oldest

and largest Balakhany-Sabunchi-Ramany field, which placed in operation in 1873-1890.

The development of corresponding deformation processes which can be identified by special studies for example monitoring of present day vertical movements of the earth crust must have been the natural consequence of the reservoir pressure decline in the fields of the Absheron peninsula. Such monitoring at the Absheron peninsula was launched as early as in 1912 when the geodynamic polygon being one of the first in the world was established here. The purpose of the first geodetic measurements was to identify differences in terms of altitude of the levels of the Caspian and Black seas.

While processing the relevelling data it was discovered that separate areas of the Absheron peninsula vary with altitude [28] . In particular, attention was paid to the subsidence of the central eastern parts of the peninsular (Figure 3). Later there was detected subsidence of the Bibi-Eybat oil field territory. For instance, LS with the speed of 30 mm/year was detected in the area of one of the oldest wells No. 2465, and in the area of the well No. 2778—31 mm/year. The similar phenomenon was observed at other fields of the peninsular (Table 1). In absolute values the presence of abnormal cases in fold areas of anticlines at the oldest oil fields in Sabunchi, Surakhany, Ramany and Bibi-Eybat reached 1 - 2.5 m for the period of 50 years, and during 80 years the central part of the Absheron peninsular subsided by more than 3 m [29] .

Though the analysis of the results of relevelling carried out on the Absheron peninsula did not contribute to unanimous opinion as to the causes of LS, however, a number of scientists link one of them to the intensive extraction of oil, gas and sand from the subsoil. Thus, according to the observations of D.A. Liliyenberg and others [30] the periods of relative tectonic LS at the fields of Binagadi, Surakhany and Bibi-Eybat correspond to the sudden increase of hydrocarbons extraction from the subsoil, whereas the periods of movement stabilization or relative elevation are associated with minimal production. The diagram shown in Figure 4 illustrates the noticeable tendency for the increase of average annual rate of the vertical earth crust movements accompanied by the increase of the average annual oil production at the fields of the Absheron peninsula (without accounting of extraction of gas and associated reservoir waters).

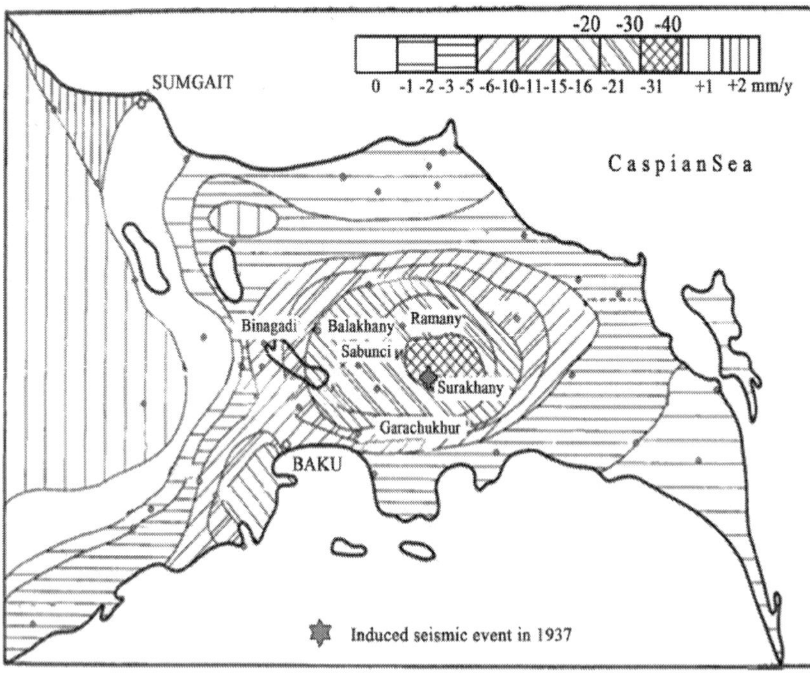

Figure 3: Scheme of vertical movements on Absheron p-la based on results repeated leveling.

Table 1: Land subsidence at some fields of the Absheron peninsular

Field	Years of levelling	Time interval, years	Annual average rate, mm/year
Binagadi	1912-1937	25	-2
Bibi-Eybat	1908-1972	64	-30
Surakhany	1926-1974	48	-30

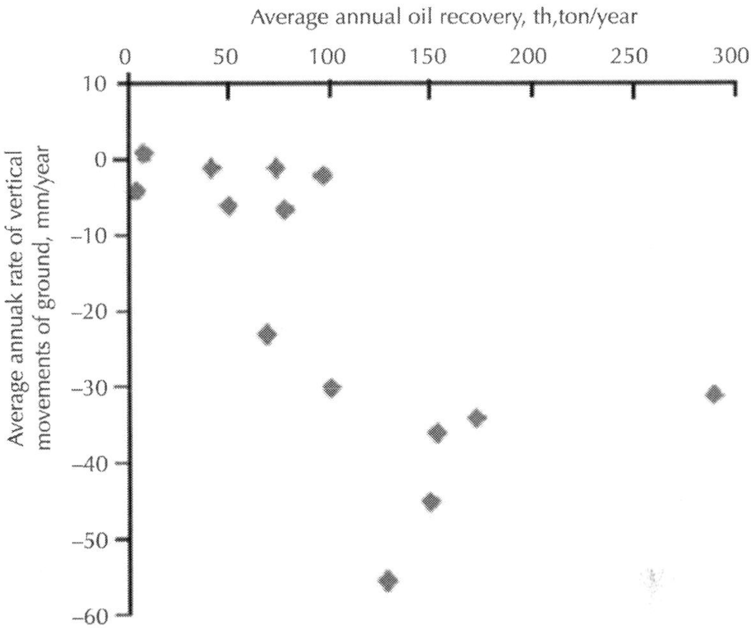

Figure 4: Diagram of dependence of earth crust movement speed on average annual oil production at the fields of Absheron peninsula (created based on Yashenko's data [28]).

The above pointed processes of land subsidence in the Absheron peninsula as well as at long term developed oil fields of other basins in the world were also accompanied by negative environmental consequences. Thus, according to the data of the hydrodynamic monitoring carried out in the Absheron peninsula there have been identified areas of intense rising of level of underground water correlating to LS areas. In particular, in the central part of the peninsular the area of zones with the depth of ground water level of >10 m decreased by 21.4% during the period from 1955 to 2006, and the area of zones with the depth of ground water level of 3 - 5 m increased by 10.3%. This serves as a vivid example of the negative impact of the long term development of oil and gas fields in this territory for hydrogeological and ecological environments which is manifested in flooding (in other cases-swamping) of territories and obvious deterioration of ameliorative properties of lands.

Abnormal rates of LS in the number of fields of the Absheron peninsula can serve as a cause of the man-in- duced seismicity. Abnormal rates of subsiding in area of Surakhany field (about 47 mm/year) have served as the reason for earthquake of magnitude 6 which occurred in 1937 near the same named village (see Figure 3).

Oil and gas production at the intensely developed oil fields of the Absheron peninsula was often accompanied by cases of borehole curving and breaking, oil, gas and water pipeline rupture, sudden water and sand kick etc. which were the results of man induced deformation processes combined with natural geodynamic processes. Thus, for example in the Balakhany-Sabunchi-Ramany field in more than 20 wells there were reported cases of breakage (wells No. 1794, 2304 and others) and breakage of borehole lines (wells No. 2409, 2575, 2610 and others) as well as their curving (wells No. 1499, 12204 and others) which led to serious economic damages [28] .

Well failures are causing large scale contamination of oil-field territories too. Two monitoring conducted in Azerbaijan over the last years by the Ministry of Ecology and Natural Resources identified around 30 thous. ha of contaminated land of which 10 thous. ha (according to other data—12 thous. ha) fall on land contaminated with oil and oil products essentially in the Absheron peninsula.

It is known that on completing the field development oil/gas wells used for the production of hydrocarbons are abandoned or temporary suspended. However, the extent of risk of dangerous (emergency) situations will not decrease as the control over their condition is often either not exercised or exercised quite cursorily without taking into account all the requirements of relevant regulatory acts [31].

According to the independent experts (no official data—Authors) in Azerbaijan only on the territory of the Absheron peninsular there is a concentration of up to 3 thousand abandoned i.e. uncontrolled oil wells. A more detailed analysis of these and a number of other environmental problems is the subject of separate researches.

CONCLUSIONS

Overview of references has shown that intensive development of oil-and-gas fields causes natural disbalance and wide progress of deformation processes: decline of reservoir pressure induced additional compaction of rocks and as a result, subsidence and displacement of a ground occur.

The above mentioned undesirable environmental consequences of the long term development of oil and gas fields in the aggregate cause considerable economic damage to the existing infrastructure. For minimization of such damage it is necessary to equip petroleum producing complexes by system of environmental control with purpose of forecasting geodynamic processes development.

REFERENCES

1. Xie, X., Jiao, J.J., Tang, Z. and Zheng, C. (2003) Evolution of Abnormally Low Pressure and Its Implications for the Hydrocarbon System in the Southeast Uplift Zone of Songliao Basin, China. AAPG Bulletin, 1, 99-119.
2. Koshlyak, V.A. (2002) Granitoid Oil and Gas Reservoirs. Tau-Press, Ufa.
3. Imanov, A.A. (2012) Physical and Mechanical Properties Sedimentary Rocks of the South-Caspian Basin in Subsurface Conditions: Deep Water Hydrocarbons Resources. Nafta-Press, Baku.
4. Morton, R.A. and Buster, N.A. (2002) Subsurface Controls on Historical Subsidence Rates and Associated Wetland Loss in South-Central Louisiana. Gulf Coast Association of Geological Societies, 52, 767-778.
5. Patel, P. and Kulkarni, M.N. (2007) Application of GPS for Monitoring Land Subsidence. Journal of Earth Sciences, 1, 35-46.

6. Serebryakov, V.A. and Chilingar, G.V. (2000) Prediction of Subsidence: Relationship between Lowering of Formation Pressure and Subsidence Due to Fluid Withdrawal. Energy Sources Part A: Recovery, Utilization and Environmental Effects, 5, 409-416.
7. dushkin, V.V. and Turuntaev, S.B. (2005) Man Induced Processes in the Earth Crust (Hazards and Catastrophes). INEK, Moscow.
8. National Research Council (1991) Mitigating Losses from Land Subsidence in the United States. National Academy Press, Washington DC.
9. Mayuga, M.N. and Allen, D.R. (1970) Subsidence in the Wilmington Oil Field, Long Beach, California, U.S.A. In: Tison, L.J., Ed., Land Subsidence, International Association of Scientific Hydrology, UNESCO, 66-79.
10. Pratt, W.E. and Johnson, D.W. (1926) Local Subsidence of the Goose Creek Oil Field. Journal of Geology, 34, 577-590. http://dx.doi.org/10.1086/623352
11. Snider, L.C. (1927) A Suggested Explanation for the Surface Subsidence in the Goose Creek Oil and Gas Field, Texas. AAPG Bulletin, 11, 729.
12. Holzer, T.L. (1990) Land Subsidence Caused by Withdrawal of Oil and Gas in the Gulf Coastal Plain, the Houston, Texas, Case History. AAPG Bulletin, 74, 1497-1498.
13. Van der Kooij, M. (1997) Land Subsidence Measurements at the Belridge Oil Fields from ERS InSAR Data. The 3rd ESA ERS Symposium, Florence.http://earth.esa.int/workshops/ers97/papers/vanderkooij1/index-2.html
14. Gabrysch, R.K. and Coplin, L.S. (1990) Land-Surface Subsidence in Houston-Galveston Region, Texas. Texas Water Development Board Report, No. 188, Austin.
15. Morton, R.A., Purcell, N.A. and Peterson, R.L. (2001) Field Evidence of Subsidence and Faulting Induced by Hydrocarbon Production in Coastal Southeast Texas. Gulf Coast Association of Geological Societies Transactions, 51, 239- 248.

16. Shinkle, K. and Dokka, R.K. (2004) Rates of Vertical Displacement at Benchmarks in the Lower Mississippi Valley and the Northern Gulf Coast. NOAA Technical Report, No. 50, 135 p.
17. Penland, S., Ramsey, K.E., McBride, R.A., Mestayer, J.T. and Westphal, K.A. (1988) Relative Sea-Level Rise and Delta-Plain Development in the Terrebonne Parish Region. Coastal Geology Tech Report, No. 4, 121 p.
18. Hejmanowski, R. (1993) Zur Vorausberechnung förderbedingter Bodensenkungen über Erdölund Erdgaslagerstätten (Prediction of Land Subsidence Induced by Oil and Gas Exploitation). Ph.D. Thesis, Clausthal University of Technology, Clausthal-Zellerfeld.
19. Melnikov, N.N. and Kalashnik, A.I. (2009) Man Induced Geodynamic Processes When Developing Oil and Gas Fields of the Barents Sea Shelf. Vestnik MGTU, 12, 601-608.
20. Nesterenko, Y.M., Kosolapov, O.V. and Nesterenko, M.Y. (2010) Seismic Activity in the Regions of the Developed Hydrocarbons Fields. News of the Samara Scientific Center of the RAS, 12, 1240-1244.
21. Kreitler, C.W., Akhter, M.S., Donnelly, A.C.A. and Wood, W.T. (1988) Hydrogeology of Formations Used for Deep-Well Injection. Texas Gulf Coast: University of Texas at Austin Bureau of Economic Geology, Austin, 204 p. Report Prepared for the U.S. Environmental Protection Agency under Cooperative Agreement No. CR812786-01-0.
22. Doornhof, D., Kristiansen, T.G., Nagel, N.B., Pattillo, P.D. and Sayers, C. (2006) Compaction and Subsidence. Oilfield Review, 18, 50-68.
23. Poland, J.F. and Davis, G.H. (1969) Land Subsidence Due to Withdrawal of Fluids. Reviews in Engineering Geology, 2, 187-269. http://dx.doi.org/10.1130/REG2-p187
24. Gorshkov, S.P. (1998) Conceptual Basics of Geoecology. Smolensk State University, Smolensk.

25. Adushkin, V.V., Rodionov, V.N. and Turuntaev, S.B. (2000) Seismicity of Hydrocarbons Fields. Oil Review, 5, 4-15.
26. Kononova, M. (2011) Earthquakes. Characteristic. Examples. St. Petersburg State University, St. Petersburg.
27. Nikolaev, A.V. (1995) Problems of Induced Seismicity. In: Nikolaev, A.V. and Galkin, I.N., Eds., Induced Seismicity, Nauka, Moscow, 5-15.
28. Yashenko, V.R. (1989) Geodesic Researches of Vertical Movements of the Earth Crust. Nedra, Moscow, 192 p.
29. Yashenko, V.R. and Yambayev, H.K. (2006) Geodesy and Eternal Mysteries of the Earth Crust Movements. Geoprofi, 4, 61-66.
30. Lilienberg, D.A., Guseinzadeh, O.D., Kuliyev, F.T., Shirinov, N.Sh. and Yashenko, V.R. (1980) Complex Studies of Present Day Tectonic Movements on the Geodynamic Polygons of Azerbaijan. In: Bulanzhe, Yu.D., Lilienberg, D.A. and Podstrigach, Ya.S., Eds., Present Day Movements of the Earth Crust. Theory, Methods, Forecast, Nauka, Moscow, 165-174.
31. Muzipov, H.N., Yerka, B.A. and Illarionova, E.G. (2009) Environmental Safety of Plants of the Exploration and Production Sector. Neftyanoe Khozaystvo, 1, 92-94.

Chapter 4

Microbial Degradation of Petroleum Hydrocarbon Contaminants: An Overview

Nilanjana Das and Preethy Chandran

Environmental Biotechnology Division, School of Biosciences and Technology, VIT University, Vellore, Tamil Nadu 632014, India

ABSTRACT

One of the major environmental problems today is hydrocarbon contamination resulting from the activities related to the petrochemical industry. Accidental releases of petroleum products are of particular concern in the environment. Hydrocarbon components have been known to belong to the family of carcinogens and neurotoxic organic pollutants. Currently accepted disposal methods of incineration or burial insecure landfills can become prohibitively expensive when amounts of contaminants are large. Mechanical and chemical methods generally used to remove hydrocarbons from contaminated sites have limited effectiveness and can be expensive. Bioremediation is the promising technology for the treatment of these contaminated sites since it is cost-effective and will lead to complete mineralization. Bioremediation functions basically on biodegradation, which may refer to complete mineralization of organic contaminants into carbon dioxide, water, inorganic compounds, and cell protein or transformation of complex organic contaminants to other simpler organic compounds by biological agents like microorganisms. Many indigenous microorganisms in water and soil are capable of degrading hydrocarbon contaminants. This paper presents an updated overview of petroleum hydrocarbon degradation by microorganisms under different ecosystems.

INTRODUCTION

Petroleum-based products are the major source of energy for industry and daily life. Leaks and accidental spills occur regularly during the exploration, production, refining, transport, and storage of petroleum and petroleum products. The amount of natural crude oil seepage was estimated to be 600,000 metric tons per year with a range of uncertainty of 200,000 metric tons per year [1]. Release of hydrocarbons into the environment whether accidentally or due to human activities is a main cause of water and soil pollution [2]. Soil contamination with hydrocarbons causes extensive damage of local system since accumulation of pollutants in animals and plant tissue may cause death or mutations [3]. The technology commonly used for the soil remediation includes mechanical, burying, evaporation, dispersion, and washing. However, these technologies are expensive and can lead to incomplete decomposition of contaminants.

The process of bioremediation, defined as the use of microorganisms to detoxify or remove pollutants owing to their diverse metabolic capabilities is an evolving method for the removal and degradation of many environmental pollutants including the products of petroleum industry [4]. In addition, bioremediation technology is believed to be noninvasive and relatively cost-effective

[5]. Biodegradation by natural populations of microorganisms represents one of the primary mechanisms by which petroleum and other hydrocarbon pollutants can be removed from the environment [6] and is cheaper than other remediation technologies [7].

The success of oil spill bioremediation depends on one's ability to establish and maintain conditions that favor enhanced oil biodegradation rates in the contaminated environment. Numerous scientific review articles have covered various factors that influence the rate of oil biodegradation [7–12]. One important requirement is the presence of microorganisms with the appropriate metabolic capabilities. If these microorganisms are present, then optimal rates of growth and hydrocarbon biodegradation can be sustained by ensuring that adequate concentrations of nutrients and oxygen are present and that the pH is between 6 and 9. The physical and chemical characteristics of the oil and oil surface area are also important determinants of bioremediation success. There are the two main approaches to oil spill bioremediation: (a) bioaugmentation, in which known oil-degrading bacteria are added to supplement the existing microbial population, and (b) biostimulation, in which the growth of indigenous oil degraders is stimulated by the addition of nutrients or other growth-limiting cosubstrates.

The success of bioremediation efforts in the cleanup of the oil tanker Exxon Valdez oil spill of 1989 [13] in Prince William Sound and the Gulf of Alaska created tremendous interest in the potential of biodegradation and bioremediation technology. Most existing studies have concentrated on evaluating the factors affecting oil bioremediation or testing favored products and methods through laboratory studies [14]. Only limited numbers of pilot scale and field trials have provided the most convincing demonstrations of this technology which have been reported in the peer-reviewed literature [15–18]. The scope of current understanding of oil bioremediation is also limited because the emphasis of most of these field studies and reviews has been given on the evaluation of bioremediation technology for dealing with large-scale oil spills on marine shorelines.

This paper provides updated information on microbial degradation of petroleum hydrocarbon contaminants towards the better understanding in bioremediation challenges.

MICROBIAL DEGRADATION OF PETROLEUM HYDROCARBONS

Biodegradation of petroleum hydrocarbons is a complex process that depends on the nature and on the amount of the hydrocarbons present. Petroleum hydrocarbons can be divided into four classes: the saturates, the aromatics, the asphaltenes (phenols, fatty acids, ketones, esters, and porphyrins), and the resins (pyridines, quinolines, carbazoles, sulfoxides, and amides) [19]. Different factors influencing hydrocarbon degradation have been reported by Cooney et al. [20]. One of the important factors that limit biodegradation of oil pollutants in the environment is their limited availability to microorganisms. Petroleum hydrocarbon compounds bind to soil components, and they are difficult to be removed or degraded [21]. Hydrocarbons differ in their susceptibility to microbial attack. The susceptibility of hydrocarbons to microbial degradation can be generally ranked as follows: linear alkanes > branched alkanes > small aromatics > cyclic alkanes [6, 22]. Some compounds, such as the high molecular weight polycyclic aromatic hydrocarbons (PAHs), may not be degraded at all [23].

Microbial degradation is the major and ultimate natural mechanism by which one can clean up the petroleum hydrocarbon pollutants from the environment [24–26]. The recognition of biodegraded petroleum-derived aromatic hydrocarbons in marine sediments was reported by Jones et al. [27]. They studied the extensive biodegradation of alkyl aromatics in marine sediments which occurred prior to detectable biodegradation of n-alkane profile of the crude oil and the microorganisms, namely, Arthrobacter, Burkholderia, Mycobacterium, Pseudomonas, Sphingomonas, and Rhodococcus were found to be involved for alkylaromatic degradation. Microbial degradation of petroleum hydrocarbons in a polluted tropical stream in Lagos, Nigeria was reported by Adebusoye et al. [28]. Nine bacterial strains, namely, Pseudomonas fluorescens, P. aeruginosa, Bacillus subtilis, Bacillus sp., Alcaligenes sp., Acinetobacter lwoffi, Flavobacterium sp., Micrococcus roseus, and Corynebacterium sp. were isolated from the polluted stream which could degrade crude oil.

Hydrocarbons in the environment are biodegraded primarily by bacteria, yeast, and fungi. The reported efficiency of biodegradation ranged from 6% [29] to 82% [30] for soil fungi, 0.13% [29] to 50% [30] for soil bacteria, and 0.003% [31] to 100% [32] for marine bacteria. Many scientists reported that mixed populations with overall broad enzymatic capacities are required to degrade complex mixtures of hydrocarbons such as crude oil in soil [33], fresh water [34], and marine environments [35, 36].

Bacteria are the most active agents in petroleum degradation, and they work as primary degraders of spilled oil in environment [37, 38]. Several bacteria are even known to feed exclusively on hydrocarbons [39]. Floodgate [36] listed 25 genera of hydrocarbon degrading bacteria and 25 genera of hydrocarbon degrading fungi which were isolated from marine environment. A similar compilation by Bartha and Bossert [33] included 22 genera of bacteria and 31 genera of fungi. In earlier days, the extent to which bacteria, yeast, and filamentous fungi participate in the biodegradation of petroleum hydrocarbons was the subject of limited study, but appeared to be a function of the ecosystem and local environmental conditions [7]. Crude petroleum oil from petroleum contaminated soil from North East India was reported by Das and Mukherjee [40]. Acinetobacter sp. was found to be capable of utilizing n-alkanes of chain length C_{10}–C_{40} as a sole source of carbon [41]. Bacterial genera, namely, Gordonia, Brevibacterium, Aeromicrobium, Dietzia, Burkholderia, and Mycobacterium isolated from petroleum contaminated soil proved to be the potential organisms for hydrocarbon degradation [42]. The degradation of poly-aromatic hydrocarbons by Sphingomonas was reported by Daugulis and McCracken [43].

Fungal genera, namely, Amorphoteca, Neosartorya, Talaromyces, and Graphium and yeast genera, namely,Candida,

Yarrowia, and Pichia were isolated from petroleum-contaminated soil and proved to be the potential organisms for hydrocarbon degradation [42]. Singh [44] also reported a group of terrestrial fungi, namely,Aspergillus, Cephalosporium, and Pencillium which were also found to be the potential degrader of crude oil hydrocarbons. The yeast species, namely, Candida lipolytica, Rhodotorula mucilaginosa, Geotrichum sp, and Trichosporon mucoides isolated from contaminated water were noted to degrade petroleum compounds [45].

Though algae and protozoa are the important members of the microbial community in both aquatic and terrestrial ecosystems, reports are scanty regarding their involvement in hydrocarbon biodegradation. Walker et al. [51] isolated an alga, Prototheca zopfi which was capable of utilizing crude oil and a mixed hydrocarbon substrate and exhibited extensive degradation of n-alkanes and isoalkanes as well as aromatic hydrocarbons. Cerniglia et al. [52] observed those nine cyanobacteria, five green algae, one red alga, one brown alga, and two diatoms could oxidize naphthalene. Protozoa, by contrast, had not been shown to utilize hydrocarbons.

FACTORS INFLUENCING PETROLEUM HYDROCARBON DEGRADATION

A number of limiting factors have been recognized to affect the biodegradation of petroleum hydrocarbons, many of which have been discussed by Brusseau [53]. The composition and inherent biodegradability of the petroleum hydrocarbon pollutant is the first and foremost important consideration when the suitability of a remediation approach is to be assessed. Among physical factors, temperature plays an important role in biodegradation of hydrocarbons by directly affecting the chemistry of the pollutants as well as affecting the physiology and diversity of the microbial flora. Atlas [54] found that at low temperatures, the viscosity of the oil increased, while the volatility of the toxic low molecular weight hydrocarbons were reduced, delaying the onset of biodegradation.

Temperature also affects the solubility of hydrocarbons [62]. Although hydrocarbon biodegradation can occur over a wide range of temperatures, the rate of biodegradation generally decreases with the decreasing temperature. Figure 1 shows that highest degradation rates that generally occur in the range 30–40°C in soil environments, 20–30°C in some freshwater environments and 15–20°C in marine environments [33, 34]. Venosa and Zhu [63] reported that ambient temperature of the environment affected both the properties of spilled oil and the activity of the microorganisms. Significant biodegradation of hydrocarbons have been reported in psychrophilic environments in temperate regions [64, 65].

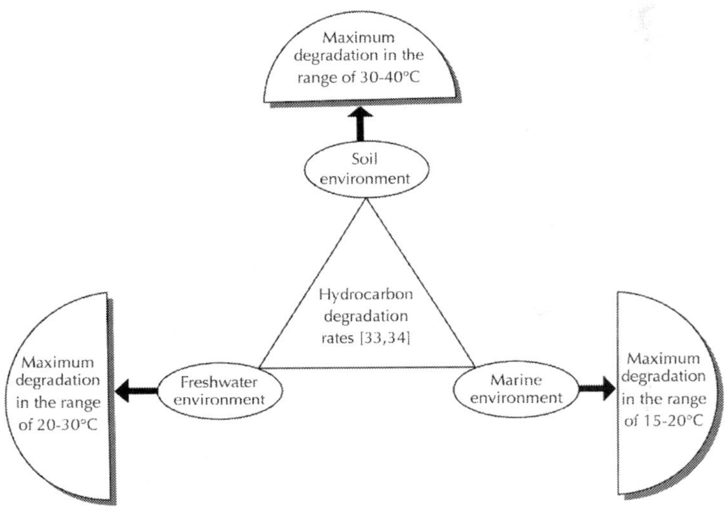

Figure 1: Hydrocarbon degradation rates in soil, fresh water, and marine environments.

Nutrients are very important ingredients for successful biodegradation of hydrocarbon pollutants especially nitrogen, phosphorus, and in some cases iron [34]. Some of these nutrients could become limiting factor thus affecting the biodegradation processes. Atlas [35] reported that when a major oil spill occurred in marine and freshwater environments, the supply of carbon was significantly increased and the availability of nitrogen and phosphorus generally became the limiting factor for oil degradation. In marine environments, it was found to be more pronounced due to low levels of nitrogen and phosphorous in seawater [36]. Freshwater wetlands are typically considered to be nutrient deficient due to heavy demands of nutrients by the plants [66]. Therefore, additions of nutrients were necessary to enhance the biodegradation of oil pollutant [67, 68]. On the other hand, excessive nutrient concentrations can also inhibit the biodegradation activity [69]. Several authors have reported the negative effects of high NPK levels on the biodegradation of hydrocarbons [70, 71] especially on aromatics [72]. The effectiveness of fertilizers for the crude oil bioremediation in subarctic intertidal sediments was studied by Pelletier et al. [64]. Use of poultry manure as organic fertilizer in contaminated soil was also reported [73], and biodegradation was found to be enhanced in the presence of poultry manure alone. Maki et al. [74] reported that photo-oxidation increased the biodegradability of petroleum hydrocarbon by increasing its bioavailability and thus enhancing microbial activities.

MECHANISM OF PETROLEUM HYDROCARBON DEGRADATION

The most rapid and complete degradation of the majority of organic pollutants is brought about under aerobic conditions. Figure 2 shows the main principle of aerobic degradation of hydrocarbons [75]. The initial intracellular attack of organic pollutants is an oxidative process and the activation as well as incorporation of oxygen is the enzymatic key reaction catalyzed by oxygenases and peroxidases. Peripheral degradation pathways convert organic pollutants step by step into intermediates of the central intermediary metabolism, for example, the tricarboxylic acid cycle. Biosynthesis of cell biomass occurs from the central precursor metabolites, for example, acetyl-

CoA, succinate, pyruvate. Sugars required for various biosyntheses and growth are synthesized by gluconeogenesis.

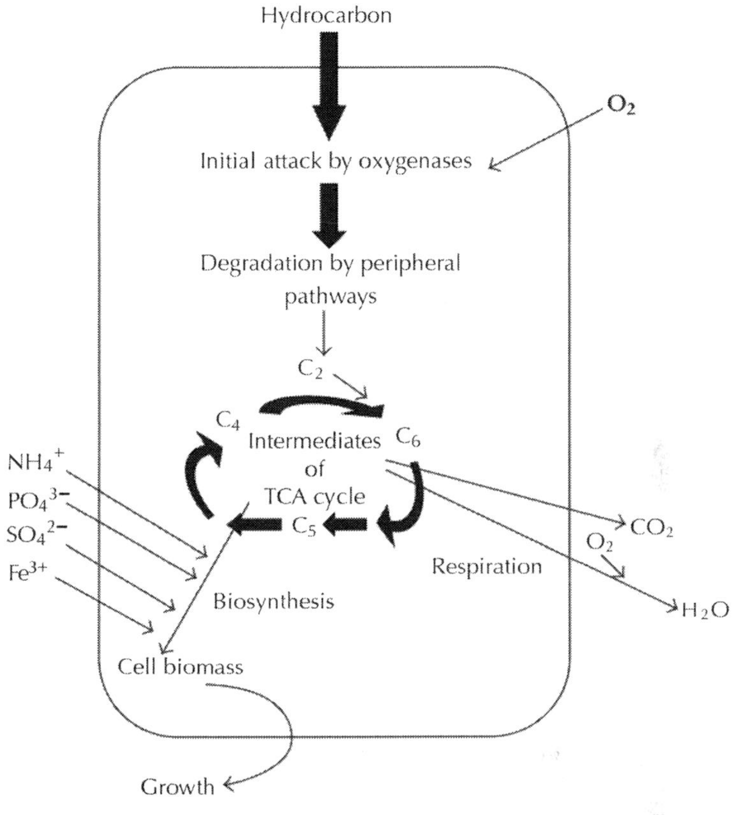

Figure 2: Main principle of aerobic degradation of hydrocarbons by microorganisms.

The degradation of petroleum hydrocarbons can be mediated by specific enzyme system. Figure 3 shows the initial attack on xenobiotics by oxygenases [75]. Other mechanisms involved are (1) attachment of microbial cells to the substrates and (2) production of biosurfactants [76]. The uptake mechanism linked to the attachment of cell to oil droplet is still unknown but production of biosurfactants has been well studied.

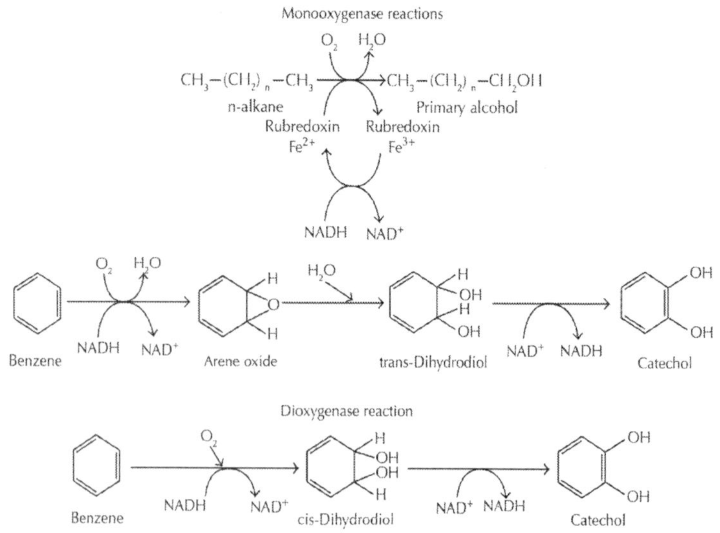

Figure 3: Enzymatic reactions involved in the processes of hydrocarbons degradation.

ENZYMES PARTICIPATING IN DEGRADATION OF HYDROCARBONS

Cytochrome P450 alkane hydroxylases constitute a super family of ubiquitous Heme-thiolate Monooxygenases which play an important role in the microbial degradation of oil, chlorinated hydrocarbons, fuel additives, and many other compounds [77]. Depending on the chain length, enzyme systems are required to introduce oxygen in the substrate to initiate biodegradation (Table 1). Higher eukaryotes generally contain several different P450 families that consist of large number of individual P450 forms that may contribute as an ensemble of isoforms to the metabolic conversion of given substrate. In microorganisms such P450 multiplicity can only be found in few species [78]. Cytochrome

P450 enzyme systems was found to be involved in biodegradation of petroleum hydrocarbons (Table 1). The capability of several yeast species to use n-alkanes and other aliphatic hydrocarbons as a sole source of carbon and energy is mediated by the existence of multiple microsomal Cytochrome P450 forms. These cytochrome P450 enzymes had been isolated from yeast species such as Candida maltosa, Candida tropicalis, and Candida apicola [79]. The diversity of alkaneoxygenase systems in prokaryotes and eukaryotes that are actively participating in the degradation of alkanes under aerobic conditions like Cytochrome P450 enzymes, integral membrane di-iron alkane hydroxylases (e.g., alkB), soluble di-iron methane monooxygenases, and membrane-bound copper containing methane monooxygenases have been discussed by Van Beilen and Funhoff [80].

Table 1: Enzymes involved in biodegradation of petroleum hydrocarbons

Enzymes	**Substrates**	**Microorganisms**	**References**
Soluble Meth-aneMonooxygen-ases	C_1–C_8 alkanes alkenes and cycloalkanes	Methylococcus	
		Methylosinus	
		Methylocystis	McDonald et al. [46]
		Methylomonas	
		Methylocella	
Particulate Meth-aneMonooxygen-ases	C_1–C_5 (halogenated) alkanes and cycloal-kanes	Methylobacter	
		Methylococcus,	McDonald et al. [46]
		Methylocystis	
AlkB relatedAl-kaneHydroxylases	C_5–C_{16} alkanes, fatty acids, alkyl benzenes, cycloalkanes and so forth	Pseudomonas	
		Burkholderia	Jan et al. [47]
		Rhodococcus,	
		Mycobacterium	
Eukaryotic P450	C_{10}–C_{16} alkanes, fatty acids	Candida maltosa	
		Candida tropicalis	Iida et al. [48]
		Yarrowia lipolytica	

Bacterial P450oxygenase system	C_5–C_{16} alkanes, cycloalkanes	Acinetobacter	
		Caulobacter	Van Beilen et al. [49]
		Mycobacterium	
Dioxygenases	C_{10}–C_{30} alkanes	Acinetobacter sp.	Maeng et al. [50]

UPTAKE OF HYDROCARBONS BY BIOSURFACTANTS

Biosurfactants are heterogeneous group of surface active chemical compounds produced by a wide variety of microorganisms [57, 58, 60, 81–83]. Surfactants enhance solubilization and removal of contaminants [84, 85]. Biodegradation is also enhanced by surfactants due to increased bioavailability of pollutants [86]. Bioremediation of oil sludge using biosurfactants has been reported by Cameotra and Singh [87]. Microbial consortium consisting of two isolates of Pseudomonas aeruginosa and one isolate Rhodococcus erythropolisfrom soil contaminated with oily sludge was used in this study. The consortium was able to degrade 90% of hydrocarbons in 6 weeks in liquid culture. The ability of the consortium to degrade sludge hydrocarbons was tested in two separate field trials. In addition, the effect of two additives (a nutrient mixture and a crude biosurfactant preparation on the efficiency of the process was also assessed. The biosurfactant used was produced by a consortium member and was identified as being a mixture of 11 rhamnolipid congeners. The consortium degraded 91% of the hydrocarbon content of soil contaminated with 1% (v/v) crude oil sludge in 5 weeks. Separate use of any one additive along with the consortium brought about a 91–95% depletion of the hydrocarbon content in 4 weeks, with the crude biosurfactant preparation being a more effective enhancer of degradation. However, more than 98% hydrocarbon depletion was obtained when both additives were added together with the consortium. The data substantiated the use of a crude biosurfactant for hydrocarbon remediation.

Pseudomonads are the best known bacteria capable of utilizing hydrocarbons as carbon and energy sources and producing biosurfactants [37, 87–89]. Among Pseudomonads, P. aeruginosa is widely studied for the production of glycolipid type biosurfactants. However, glycolipid type biosurfactants are also reported from some other species like P. putida and P. chlororaphis. Biosurfactants increase the oil surface area and that amount of oil is actually available for bacteria to utilize it [90]. Table 2 summarizes the recent reports on biosurfactant production by different microorganisms. Biosurfactants can act as emulsifying agents by decreasing the surface tension and forming micelles. The microdroplets encapsulated in the hydrophobic microbial cell surface are taken inside and degraded. Figure 4 demonstrates the involvement of biosurfactant (rhamnolipids) produced by Pseudomonas sp. and the mechanism of formation of micelles in the uptake of hydrocarbons [75].

Table 2: Biosurfactants produced by microorganisms

Biosurfactants	**Microorganisms**
Sophorolipids	Candida bombicola (Daverey and Pakshirajan, [55])
Rhamnolipids	Pseudomonas aeruginosa (Kumar et al. [56])
Lipomannan	Candida tropicalis (Muthuswamy et al. [57])
Rhamnolipids	Pseudomonas fluorescens (Mahmound et al. [58])
Surfactin	Bacillus subtilis (Youssef et al. [59])
Glycolipid	Aeromonas sp. (Ilori et al. [60])
Glycolipid	Bacillus sp. (Tabatabaee et al. [61])

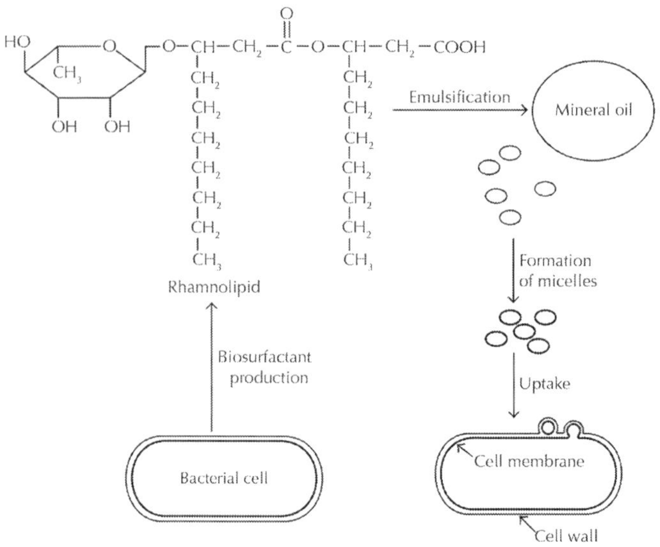

Figure 4: Involvement of biosurfactant (rhamnolipid) produced by Pseudomonas sp in the uptake of hydrocarbons.

BIODEGRADATION OF PETROLEUM HYDROCARBONS BY IMMOBILIZED CELLS

Immobilized cells have been used and studied for the bioremediation of numerous toxic chemicals. Immobilization not only simplifies separation and recovery of immobilized cells but also makes the application reusable which reduces the overall cost. Wilsey and Bradely [91] used free suspension and immobilizedPseudomonas sp. to degrade petrol in an aqueous system. The study indicated that immobilization resulted in a combination of increased contact between cell and hydrocarbon droplets and enhanced level of rhamnolipids production. Rhamnolipids caused greater dispersion of water-insoluble n-alkanes in the aqueous phase due to their amphipathic properties and the molecules consist of hydrophilic and hydrophobic moieties reduced the interfacial tension of oil-water systems. This resulted in higher interaction of cells with solubilized hydrocarbon droplets much smaller than the cells and rapid uptake of hydrocarbon in to the cells. Diaz et al. [92] reported that immobilization of bacterial cells enhanced the biodegradation rate of crude oil compared to free living cells in a wide range of culture salinity. Immobilization can be done in batch mode as well as continuous mode. Packed bed reactors are commonly used in continuous mode to degrade hydrocarbons. Cunningham et al. [93] used polyvinyl alcohol (PVA) cryogelation as an entrapment matrix and microorganisms indigenous to the site. They constructed laboratory biopiles to compare immobilised bioaugmentation with liquid culture bioaugmentation and biostimulation. Immobilised systems were found to be the most successful in terms of percentage removal of diesel after 32 days.

Rahman et al. [94] conducted an experiment to study the capacity of immobilized bacteria in alginate beads to degrade hydrocarbons. The results showed that there was no decline in the biodegradation activity of the microbial consortium on the repeated use. It was concluded that immobilization of cells are a promising application in the bioremediation of hydrocarbon contaminated site.

COMMERCIALLY AVAILABLE BIOREMEDIATION AGENTS

Microbiological cultures, enzyme additives, or nutrient additives that significantly increase the rate of biodegradation to mitigate the effects of the discharge were defied as bioremediation agents by U.S.EPA [95]. Bioremediation agents are classified as bioaugmentation agents and biostimulation agents based on the two main approaches to oil spill bioremediation. Numerous bioremediation products have been proposed and promoted by their vendors, especially during early 1990s, when bioremediation was popularized as "the ultimate solution" to oil spills [96].

The U.S. EPA compiled a list of 15 bioremediation agents [95, 97] as a part of the National Oil and Hazardous Substances Pollution Contingency Plan (NCP) Product Schedule, which was required by the Clean Water Act, the Oil Pollution Act of 1990, and the National Contingency Plan (NCP) as shown in Table 3. But the list was modified, and the number of bioremediation agents was reduced to nine.

Table 3: Bioremediation agents in NCP product schedule (Adapted from USEPA, 2002)

Name or Trademark	Product Type	Manufacture
BET BIOPETRO	MC	BioEnviro Tech, Tomball, TX
BILGEPRO	NA	International Environmental Products, LLC, Conshohocken, PA.
INIPOL EAP 22	NA	Societe, CECA S.A., France
LAND AND SEA	NA	Land and Sea Restoration LLC, San Antonio, TX
RESTORATION MICRO-BLAZE	MC	Verde Environmental, Inc., Houston, TX
OIL SPILL EATER II	NA/EA	Oil Spill Eater International, Corporation, Dallas, TX
OPPENHEIMER FORMULA	MC	Oppenheimer Biotechnology, Inc., Austin, TX
PRISTINE SEA II	MC	Marine Systems, Baton Rouge, LA
STEP ONE	MC	B & S Research, Inc., Embarrass, MN
SYSTEM E.T. 20.	MC	Quantum Environmental Technologies, Inc(QET), La Jolla, CA
VB591TMWATER, VB997TMSOIL, AND BINUTRIX	NA	BioNutraTech, Inc., Houston,TX
WMI-2000	MC	WMI International, Inc

Abbreviations of product type:

MC: Microbial Culture

EA: Enzyme Additive

NA: Nutrient Additive.

Studies showed that bioremediation products may be effective in the laboratory but significantly less so in the field [14, 17, 18, 98]. This is because laboratory studies cannot always simulate complicated real world conditions such as spatial heterogeneity, biological interactions, climatic effects, and nutrient mass transport limitations. Therefore, field studies and applications are the ultimate tests or the most convincing demonstration of the effectiveness of bioremediation products.

Compared to microbial products, very few nutrient additives have been developed and marketed specifically as commercial bioremediation agents for oil spill cleanup. It is probably because common fertilizers are inexpensive, readily available, and has been shown effective if used properly. However, due to the limitations of common fertilizers (e.g., being rapidly washed out due to tide and wave action), several organic nutrient products, such as oleophilic nutrient products, have recently been evaluated and marketed as bioremediation agents. Four agents, namely, Inipol EAP22, Oil Spill Eater II (OSE II), BIOREN 1, and BIOREN 2, listed on the NCP Product Schedule have also been put into this category.

Inipol EAP22 (Societe, CECA S.A., France) is listed on the NCP Product Schedule as a nutrient additive and probably the most well-known bioremediation agent for oil spill cleanup due to its use in Prince William Sound, Alaska. This nutrient product is a microemulsion-containing urea as a nitrogen source, sodium laureth phosphate as a phosphorus source, 2-butoxy-1-ethanol as a surfactant, and oleic acid to give the material its hydrophobicity. The claimed advantages of Inipol EAP22 include (1) preventing the formation of water-in-oil emulsions by reducing the oil viscosity and interfacial tension; (2) providing controlled release of nitrogen and phosphorus for oil biodegradation; (3) exhibiting no toxicity to flora and fauna and good biodegradability [99].

Oil Spill Eater II (Oil Spill Eater International, Corp.) is another nutrient product listed on the NCP Schedule [97]. This product is listed as a nutrient/enzyme additive and consists of "nitrogen, phosphorus, readily available carbon, and vitamins for quick colonization of naturally occurring bacteria". A field demonstration was carried out at a bioventing site in a Marine Corps Air Ground Combat Center (MCAGCC) in California to investigate the efficacy of OSEII for enhancing hydrocarbon biodegradation in a fuel-contaminated vadose zone [106].

Researchers from European EUREKA BIOREN program conducted a field trial in an estuary environment to evaluate the effectiveness of two bioremediation products (BIOREN 1 and 2) [114, 115]. The two nutrient products were derived from fish meals in a granular form with urea and super phosphate as nitrogen and phosphorus sources and proteinaceous material as the carbon source. The major difference between the two formulations was that BIOREN 1 contained a biosurfactant. The results showed that the presence of biosurfactant in BIOREN 1 was the most active ingredient which contributed to the increase in oil degradation rates whereas BIOREN 2 (without biosurfactant) was not effective in that respect. The biosurfactant could have contributed to greater bioavailability of hydrocarbons to microbial attack.

PHYTOREMEDIATION

Phytoremediation is an emerging technology that uses plants to manage a wide variety of environmental pollution problems, including the cleanup of soils and groundwater contaminated with hydrocarbons and other hazardous substances. The different mechanisms, namely, hydraulic control, phytovolatilization, rhizoremediation, and phytotransformation could be utilized for the remediation of a wide variety of contaminants.

Phytoremediation can be cost-effective (a) for large sites with shallow residual levels of contamination by organic, nutrient, or metal pollutants, where contamination does not pose an imminent danger and only "polishing treatment" is required; (b) where vegetation is used as a final cap and closure of the site [116].

Advantages of using phytoremediation include cost-effectiveness, aesthetic advantages, and long-term applicability

(Table 4). Furthermore, the use of phytoremediation as a secondary or polishing in situ treatment step minimizes land disturbance and eliminates transportation and liability costs associated with offsite treatment and disposal.

Table 4: Advantages and disadvantages of phytoremediation over traditional technologies

Advantages	**Disadvantages**
Relatively low cost	Longer remediation times
Easily implemented and maintained	Climate dependent
Several mechanisms for removal	Effects to food web might be unknown
Environmentally friendly	Ultimate contaminant fates might be unknown
Aesthetically pleasing	Results are variable
Reduces landfilled wastes	
Harvestable plant material	

Research and application of phytoremediation for the treatment of petroleum hydrocarbon contamination over the past fifteen years have provided much useful information that can be used to design effective remediation systems and drive further improvement and innovation. Phytoremediation could be applied for the remediation of numerous contaminated sites. However, not much is known about contaminant fate and transformation pathways, including the identity of metabolites (Table 4). Little data exists on contaminant removal rates and efficiencies directly attributable to plants under field conditions.

The potential use of phytoremediation at a site contaminated with hydrocarbons was investigated. The Alabama Department of Environmental Management granted a site, which involved about 1500 cubic yards of soil of which 70% of the baseline samples contained over 100 ppm of total petroleum hydrocarbon (TPH). After 1 year of vegetative cover, approximately 83% of the samples were found to contain less than 10-ppm TPH. Removal of total petroleum hydrocarbon (TPH) at several field sites contaminated with crude oil, diesel fuel, or petroleum refinery wastes, at initial TPH concentrations of 1,700 to 16,000 mg/kg were also investigated [117, 118]. Plant growth was found to vary depending upon the species. Presence of some species led to greater TPH disappearance than with other species or in unvegetated soil. Among tropical plants tested for use in Pacific Islands, three coastal trees, kou (Cordia subcordata), milo (Thespesia populnea), and kiawe (Prosopis pallida) and the native shrub beach naupaka (Scaevola serica) tolerated field conditions and facilitated cleanup of soils contaminated with diesel fuel [119]. Grasses were often planted with trees at sites with organic contaminants as the primary remediation method. Tremendous amount of fine roots in the surface soil was found to be effective at binding and transforming hydrophobic contaminants such as TPH, BTEX, and PAHs. Grasses were often planted between rows of trees to provide soil stabilization and protection against wind-blown dust that could move contaminants offsite. Legumes such as alfalfa (Medicago sativa), alsike clover (Trifolium hybridum), and peas (Pisum sp.) could be used to restore nitrogen to poor soils. Fescue (Vulpia myuros), rye (Elymus sp.), clover (Trifolium sp.), and reed canary grass (Phalaris arundinacea) were used successfully at several sites, especially contaminated with petrochemical wastes. Once harvested, the grasses could be disposed of as compost or burned. Microbial degradation in the rhizosphere might be the most significant mechanism for removal of diesel range organics in vegetated contaminated soils [120]. This occurs because contaminants such as PAHs are highly hydrophobic, and their sorption to soil decreases their bioavailability for plant uptake and phytotransformation.

GENETICALLY MODIFIED BACTERIA

Applications for genetically engineered microorganisms (GEMs) in bioremediation have received a great deal of attention to improve the degradation of hazardous wastes under laboratory conditions. There are reports on the degradation of environmental pollutants by different bacteria. Table 5 shows some examples of the relevant use of genetic engineering technology to improve bioremediation of hydrocarbon contaminants using bacteria. The genetically engineered bacteria showed higher degradative capacity. However, ecological and environmental concerns and regulatory constraints are major obstacles for testing GEM in the field. These problems must be solved before GEM can provide an effective clean-up process at lower cost.

Table 5: Genetic engineering for biodegradation of contaminants

Microorganisms	Modification	Contaminants	Reference
Pseudomonas. putida	pathway	4-ethylbenzoate	Ramos et al. [100]
P. putida KT2442	pathway	toluene/benzoate	Panke and Sanchezromero [101]
Pseudomonas sp.FRI	pathway	chloro-, methylbenzoates	Rojo et al. [102]
Comamonas. Testosterone VP44	substrate specificity	o-, p-monochlorobiphenyls	Hrywna et al. [103]
Pseudomonas sp. LB400	substrate specificity	PCB	Erickson and Mondello [104]
P. pseudoalcaligenes KF707-D2	substrate specificity	TCE, toluene, benzene	Suyama et al. [105]

The use of genetically engineered bacteria was applied to bioremediation process monitoring, strain monitoring, stress response, end-point analysis, and toxicity assessment. Examples of these applications are listed in Table 6. The range of tested contaminants included chlorinated compounds, aromatic hydrocarbons, and nonpolar toxicants. The combination of microbiological and ecological knowledge, biochemical mechanisms, and field engineering designs are essential elements for successful in situ bioremediation using genetically modified bacteria.

Table 6: Application of genetically modified bacteria for assessing the biodegradation process efficiency

Microorganisms	Application	Contaminants	Reference
A. eutrophus H850Lr	process monitoring	PCB	Van Dyke et al. [107]
P. putida TVA8	process monitoring	TCE, BTEX	Applegate et al. [108]
P. fluorescens HK44	process monitoring	naphthalene, anthracene	Sayler and Ripp [109]
B. cepacia BRI6001L	strain monitoring	2,4-D	Masson et al. [110]
P. fluorescens 10586s/pUCD607	stress response	BTEX	Sousa et al. [111]
Pseudomonas strain Shk1	toxicity assessment	2, 4-dinitrophenol hydroquinone	Kelly et al. [112]
A. eutrophus 2050	end point analysis	non polar narcotics	Layton et al. [113]

CONCLUSIONS

Cleaning up of petroleum hydrocarbons in the subsurface environment is a real world problem. A better understanding of the mechanism of biodegradation has a high ecological significance that depends on the indigenous microorganisms to transform or mineralize the organic contaminants. Microbial degradation process aids the elimination of spilled oil from the environment after critical removal of large amounts of the oil by various physical and chemical methods. This is possible because microorganisms have enzyme systems to degrade and utilize different hydrocarbons as a source of carbon and energy.

The use of genetically modified (GM) bacteria represents a research frontier with broad implications. The potential benefits of using genetically modified bacteria are significant. But the need for GM bacteria may be questionable for many cases, considering that indigenous species often perform adequately but we do not tap the full potential of wild species due to our limited understanding of various phytoremediation mechanisms, including the regulation of enzyme systems that degrade pollutants.

Therefore, based on the present review, it may be concluded that microbial degradation can be considered as a key component in the cleanup strategy for petroleum hydrocarbon remediation.

REFERENCES

1. K. A. Kvenvolden and C. K. Cooper, "Natural seepage of crude oil into the marine environment," Geo-Marine Letters, vol. 23, no. 3-4, pp. 140–146, 2003.
2. C. Holliger, S. Gaspard, G. Glod, C. Heijman, W. Schumacher, R. P. Schwarzenbach, and F. Vazquez, "Contaminated environments in the subsurface and bioremediation: organic contaminants," FEMS Microbiology Reviews, vol. 20, no. 3-4, pp. 517–523, 1997.
3. P. J. J. Alvarez and T. M. Vogel, "Substrate interactions of benzene, toluene, and para-xylene during microbial degradation by pure cultures and mixed culture aquifer slurries," Applied and Environmental Microbiology, vol. 57, no. 10, pp. 2981–2985, 1991.
4. J. I. Medina-Bellver, P. Marín, A. Delgado, A. Rodríguez-Sánchez, E. Reyes, J. L. Ramos, and S. Marqués, "Evidence for in situ crude oil biodegradation after the Prestige oil spill," Environmental Microbiology, vol. 7, no. 6, pp. 773–779, 2005.
5. T. M. April, J. M. Foght, and R. S. Currah, "Hydrocarbon-degrading filamentous fungi isolated from flare pit soils in northern and western Canada," Canadian Journal of Microbiology, vol. 46, no. 1, pp. 38–49, 2000.
6. W. Ulrici, "Contaminant soil areas, different countries and contaminant monitoring of contaminants," in Environmental

Process II. Soil Decontamination Biotechnology, H. J. Rehm and G. Reed, Eds., vol. 11, pp. 5–42, 2000.

7. J. G. Leahy and R. R. Colwell, "Microbial degradation of hydrocarbons in the environment,"Microbiological Reviews, vol. 54, no. 3, pp. 305–315, 1990.
8. C. E. Zobell, "Action of microorganisms on hydrocarbons," Bacteriological Reviews, vol. 10, pp. 1–49, 1946.
9. R.M.Atlas,"Microbial degradation of petroleum hydrocarbons: an environmental perspective,"Microbiological Reviews, vol. 45, no. 1, pp. 180–209, 1981.
10. R. M. Atlas, Ed., Petroleum Microbiology, Macmillion, New York, NY, USA, 1984.
11. R. M. Atlas and R. Bartha, "Hydrocabon biodegradation and oil spill bioremediation," Advances in Microbial Ecology, vol. 12, pp. 287–338, 1992.
12. J. M. Foght and D. W. S. Westlake, "Biodegradation of hydrocarbons in freshwater," in Oil in Freshwater: Chemistry, Biology, Countermeasure Technology, J. H. Vandermeulen and S. R. Hrudey, Eds., pp. 217–230, Pergamon Press, New York, NY, USA, 1987.
13. R. M. Atlas and R. Bartha, "Fundamentals and applications," in Microbial Ecology, pp. 523–530, Benjamin/Cummings, San Francisco, Calif, USA, 4th edition, 1998.
14. A. J. Mearns, "Cleaning oiled shores: putting bioremediation to the test," Spill Science and Technology Bulletin, vol. 4, no. 4, pp. 209–217, 1997.
15. R. C. Prince, "Petroleum spill bioremediation in marine environments," Critical Reviews in Microbiology, vol. 19, no. 4, pp. 217–242, 1993.
16. R. P. J. Swannell, K. Lee, and M. Mcdonagh, "Field evaluations of marine oil spill bioremediation,"Microbiological Reviews, vol. 60, no. 2, pp. 342–365, 1996.
17. A. D. Venosa, M. T. Suidan, and M. T. Suidan, "Bioremediation of an experimental oil spill on the shoreline of Delaware Bay,"

Environmental Science and Technology, vol. 30, no. 5, pp. 1764–1775, 1996.

18. A. D. Venosa, D. W. King, and G. A. Sorial, "The baffled flask test for dispersant effectiveness: a round Robin evaluation of reproducibility and repeatability," Spill Science and Technology Bulletin, vol. 7, no. 5-6, pp. 299–308, 2002.
19. R. R. Colwell, J. D. Walker, and J. J. Cooney, "Ecological aspects of microbial degradation of petroleum in the marine environment," Critical Reviews in Microbiology, vol. 5, no. 4, pp. 423–445, 1977.
20. J. J. Cooney, S. A. Silver, and E. A. Beck, "Factors influencing hydrocarbon degradation in three freshwater lakes," Microbial Ecology, vol. 11, no. 2, pp. 127–137, 1985.
21. S. Barathi and N. Vasudevan, "Utilization of petroleum hydrocarbons by Pseudomonas fluorescensisolated from a petroleum-contaminated soil," Environment International, vol. 26, no. 5-6, pp. 413–416, 2001.
22. J. J. Perry, "Microbial metabolism of cyclic alkanes," in Petroleum Microbiology, R. M. Atlas, Ed., pp. 61–98, Macmillan, New York, NY, USA, 1984.
23. R. Atlas and J. Bragg, "Bioremediation of marine oil spills: when and when not—the Exxon Valdez experience," Microbial Biotechnology, vol. 2, no. 2, pp. 213–221, 2009.
24. R. M. Atlas, "Petroleum microbiology," in Encyclopedia of Microbiology, pp. 363–369, Academic Press, Baltimore, Md, USA, 1992.
25. O. O. Amund and N. Nwokoye, "Hydrocarbon potentials of yeast isolates from a polluted Lagoon,"Journal of Scientific Research and Development, vol. 1, pp. 65–68, 1993.
26. B. Lal and S. Khanna, "Degradation of crude oil by Acinetobacter calcoaceticus andAlcaligenes odorans,"Journal of Applied Bacteriology, vol. 81, no. 4, pp. 355–362, 1996.
27. D. M. Jones, A. G. Douglas, R. J. Parkes, J. Taylor, W. Giger, and C. Schaffner, "The recognition of biodegraded petroleum-

derived aromatic hydrocarbons in recent marine sediments," Marine Pollution Bulletin, vol. 14, no. 3, pp. 103–108, 1983.

28. S. A. Adebusoye, M. O. Ilori, O. O. Amund, O. D. Teniola, and S. O. Olatope, "Microbial degradation of petroleum hydrocarbons in a polluted tropical stream," World Journal of Microbiology and Biotechnology, vol. 23, no. 8, pp. 1149–1159, 2007.
29. J. Jones, M. Knight, and J. A. Byron, "Effect of gross population by kerosene hydrocarbons on the microflora of a moorland soil," Nature, vol. 227, p. 1166, 1970.
30. Y. Pinholt, S. Struwe, and A. Kjoller, "Microbial changes during oil decomposition in soil," Holarctic Ecology, vol. 2, pp. 195–200, 1979.
31. S. L. Hollaway, G. M. Faw, and R. K. Sizemore, "The bacterial community composition of an active oil field in the Northwestern Gulf of Mexico," Marine Pollution Bulletin, vol. 11, no. 6, pp. 153–156, 1980.
32. G. J. Mulkins Phillips and J. E. Stewart, "Distribution of hydrocarbon utilizing bacteria in Northwestern Atlantic waters and coastal sediments," Canadian Journal of Microbiology, vol. 20, no. 7, pp. 955–962, 1974.
33. R. Bartha and I. Bossert, "The treatment and disposal of petroleum wastes," in Petroleum Microbiology, R. M. Atlas, Ed., pp. 553–578, Macmillan, New York, NY, USA, 1984.
34. J. J. Cooney, "The fate of petroleum pollutants in fresh water ecosystems," in Petroleum Microbiology, R. M. Atlas, Ed., pp. 399–434, Macmillan, New York, NY, USA, 1984.
35. R. M. Atlas, "Effects of hydrocarbons on micro-organisms and biodegradation in Arctic ecosystems," inPetroleum Effects in the Arctic Environment, F. R. Engelhardt, Ed., pp. 63–99, Elsevier, London, UK, 1985.
36. G. Floodgate, "The fate of petroleum in marine ecosystems," in Petroleum Microbiology, R. M. Atlas, Ed., pp. 355–398, Macmillion, New York, NY, USA, 1984.

37. K. S. M. Rahman, T. J. Rahman, Y. Kourkoutas, I. Petsas, R. Marchant, and I. M. Banat, "Enhanced bioremediation of n-alkane in petroleum sludge using bacterial consortium amended with rhamnolipid and micronutrients," Bioresource Technology, vol. 90, no. 2, pp. 159–168, 2003.
38. R. J. W. Brooijmans, M. I. Pastink, and R. J. Siezen, "Hydrocarbon-degrading bacteria: the oil-spill clean-up crew," Microbial Biotechnology, vol. 2, no. 6, pp. 587–594, 2009.
39. M. M. Yakimov, K. N. Timmis, and P. N. Golyshin, "Obligate oil-degrading marine bacteria," Current Opinion in Biotechnology, vol. 18, no. 3, pp. 257–266, 2007.
40. K. Das and A. K. Mukherjee, "Crude petroleum-oil biodegradation efficiency of Bacillus subtilis andPseudomonas aeruginosa strains isolated from a petroleum-oil contaminated soil from North-East India," Bioresource Technology, vol. 98, no. 7, pp. 1339–1345, 2007. ·
41. M. Throne-Holst, A. Wentzel, T. E. Ellingsen, H.-K. Kotlar, and S. B. Zotchev, "Identification of novel genes involved in long-chain n-alkane degradation by Acinetobacter sp. strain DSM 17874," Applied and Environmental Microbiology, vol. 73, no. 10, pp. 3327–3332, 2007. ·
42. F. Chaillan, A. Le Flèche, E. Bury, Y.-H. Phantavong, P. Grimont, A. Saliot, and J. Oudot, "Identification and biodegradation potential of tropical aerobic hydrocarbon-degrading microorganisms," Research in Microbiology, vol. 155, no. 7, pp. 587–595, 2004. ·
43. A. J. Daugulis and C. M. McCracken, "Microbial degradation of high and low molecular weight polyaromatic hydrocarbons in a two-phase partitioning bioreactor by two strains of Sphingomonas sp,"Biotechnology Letters, vol. 25, no. 17, pp. 1441–1444, 2003.
44. H. Singh, Mycoremediation: Fungal Bioremediation, Wiley-Interscience, New York, NY, USA, 2006.

45. E. Bogusławska-Was and W. Dąbrowski, "The seasonal variability of yeasts and yeast-like organisms in water and bottom sediment of the Szczecin Lagoon," International Journal of Hygiene and Environmental Health, vol. 203, no. 5-6, pp. 451–458, 2001.

46. I. R. McDonald, C. B. Miguez, G. Rogge, D. Bourque, K. D. Wendlandt, D. Groleau, and J. Murrell, "Diversity of soluble methane monooxygenase-containing methanotrophs isolated from polluted environments," FEMS Microbiology Letters, vol. 255, no. 2, pp. 225–232, 2006.

47. B. Jan, V. Beilen, M. Neuenschwunder, T. H. M. Suits, C. Roth, S. B. Balada, and B. Witholt, "Rubredoxins involved in alkane degradation," The Journal of Bacteriology, vol. 184, no. 6, pp. 1722–1732, 2003.

48. T. Iida, T. Sumita, A. Ohta, and M. Takagi, "The cytochrome P450ALK multigene family of an n-alkane-assimilating yeast, Yarrowia lipolytica: cloning and characterization of genes coding for new CYP52 family members," Yeast, vol. 16, no. 12, pp. 1077–1087, 2000.

49. J. B. Van Beilen, E. G. Funhoff, and E. G. Funhoff, "Cytochrome P450 alkane hydroxylases of the CYP153 family are common in alkane-degrading eubacteria lacking integral membrane alkane hydroxylases," Applied and Environmental Microbiology, vol. 72, no. 1, pp. 59–65, 2006.

50. J. H. O. Maeng, Y. Sakai, Y. Tani, and N. Kato, "Isolation and characterization of a novel oxygenase that catalyzes the first step of n-alkane oxidation in Acinetobacter sp. strain M-1," Journal of Bacteriology, vol. 178, no. 13, pp. 3695–3700, 1996.

51. J. D. Walker, R. R. Colwell, Z. Vaituzis, and S. A. Meyer, "Petroleum degrading achlorophyllous algaPrototheca zopfii," Nature, vol. 254, no. 5499, pp. 423–424, 1975.

52. C. E. Cerniglia, D. T. Gibson, and C. Van Baalen, "Oxidation of naphthalene by cyanobacteria and microalgae," Journal of General Microbiology, vol. 116, no. 2, pp. 495–500, 1980.

53. M. L. Brusseau, "The impact of physical, chemical and biological factors on biodegradation," inProceedings of the International Conference on Biotechnology for Soil Remediation: Scientific Bases and Practical Applications, R. Serra, Ed., pp. 81–98, C.I.P.A. S.R.L., Milan, Italy, 1998.
54. R. M. Atlas, "Effects of temperature and crude oil composition on petroleum biodegradation," Journal of Applied Microbiology, vol. 30, no. 3, pp. 396–403, 1975.
55. A. Daverey and K. Pakshirajan, "Production of sophorolipids by the yeast Candida bombicola using simple and low cost fermentative media," Food Research International, vol. 42, no. 4, pp. 499–504, 2009.
56. M. Kumar, V. León, A. De Sisto Materano, O. A. Ilzins, and L. Luis, "Biosurfactant production and hydrocarbon-degradation by halotolerant and thermotolerant Pseudomonas sp," World Journal of Microbiology and Biotechnology, vol. 24, no. 7, pp. 1047–1057, 2008.
57. K. Muthusamy, S. Gopalakrishnan, T. K. Ravi, and P. Sivachidambaram, "Biosurfactants: properties, commercial production and application," Current Science, vol. 94, no. 6, pp. 736–747, 2008.
58. A. Mahmound, Y. Aziza, A. Abdeltif, and M. Rachida, "Biosurfactant production by Bacillus strain injected in the petroleum reservoirs," Journal of Industrial Microbiology & Biotechnology, vol. 35, pp. 1303–1306, 2008.
59. N. Youssef, D. R. Simpson, K. E. Duncan, M. J. McInerney, M. Folmsbee, T. Fincher, and R. M. Knapp, "In situ biosurfactant production by Bacillus strains injected into a limestone petroleum reservoir,"Applied and Environmental Microbiology, vol. 73, no. 4, pp. 1239–1247, 2007.
60. M. O. Ilori, C. J. Amobi, and A. C. Odocha, "Factors affecting biosurfactant production by oil degradingAeromonas spp. isolated from a tropical environment," Chemosphere, vol. 61, no. 7, pp. 985–992, 2005.

61. A. Tabatabaee, M. M. Assadi, A. A. Noohi, and V. A. Sajadian, "Isolation of biosurfactant producing bacteria from oil reservoirs," Iranian Journal of Environmental Health Science & Engineering, vol. 2, no. 1, pp. 6–12, 2005.
62. J. M. Foght, D. W. S. Westlake, W. M. Johnson, and H. F. Ridgway, "Environmental gasoline-utilizing isolates and clinical isolates of Pseudomonas aeruginosa are taxonomically indistinguishable by chemotaxonomic and molecular techniques," Microbiology, vol. 142, no. 9, pp. 2333–2340, 1996.
63. A. D. Venosa and X. Zhu, "Biodegradation of crude oil contaminating marine shorelines and freshwater wetlands," Spill Science and Technology Bulletin, vol. 8, no. 2, pp. 163–178, 2003.
64. E. Pelletier, D. Delille, and B. Delille, "Crude oil bioremediation in sub-Antarctic intertidal sediments: chemistry and toxicity of oiled residues," Marine Environmental Research, vol. 57, no. 4, pp. 311–327, 2004.
65. D. Delille, F. Coulon, and E. Pelletier, "Effects of temperature warming during a bioremediation study of natural and nutrient-amended hydrocarbon-contaminated sub-Antarctic soils," Cold Regions Science and Technology, vol. 40, no. 1-2, pp. 61–70, 2004.
66. W. J. Mitsch and J. G. Gosselink, Wetlands, John Wiley & Sons, New York, NY, USA, 2nd edition, 1993.
67. S.-C. Choi, K. K. Kwon, J. H. Sohn, and S.-J. Kim, "Evaluation of fertilizer additions to stimulate oil biodegradation in sand seashore mesocosms," Journal of Microbiology and Biotechnology, vol. 12, no. 3, pp. 431–436, 2002.
68. S.-J. Kim, D. H. Choi, D. S. Sim, and Y.-S. Oh, "Evaluation of bioremediation effectiveness on crude oil-contaminated sand," Chemosphere, vol. 59, no. 6, pp. 845–852, 2005.
69. F. Chaillan, C. H. Chaîneau, V. Point, A. Saliot, and J. Oudot, "Factors inhibiting bioremediation of soil contaminated with

weathered oils and drill cuttings," Environmental Pollution, vol. 144, no. 1, pp. 255–265, 2006.

70. J. Oudot, F. X. Merlin, and P. Pinvidic, "Weathering rates of oil components in a bioremediation experiment in estuarine sediments," Marine Environmental Research, vol. 45, no. 2, pp. 113–125, 1998.
71. C. H. Chaîneau, G. Rougeux, C. Yéprémian, and J. Oudot, "Effects of nutrient concentration on the biodegradation of crude oil and associated microbial populations in the soil," Soil Biology and Biochemistry, vol. 37, no. 8, pp. 1490–1497, 2005.
72. L. M. Carmichael and F. K. Pfaender, "The effect of inorganic and organic supplements on the microbial degradation of phenanthrene and pyrene in soils," Biodegradation, vol. 8, no. 1, pp. 1–13, 1997.
73. J. C. Okolo, E. N. Amadi, and C. T. I. Odu, "Effects of soil treatments containing poultry manure on crude oil degradation in a sandy loam soil," Applied Ecology and Environmental Research, vol. 3, no. 1, pp. 47–53, 2005.
74. H. Maki, T. Sasaki, and S. Haramaya, "Photooxidation of biodegradable crude oil and toxicity of the photooxidized products," Chemosphere, vol. 44, pp. 1145–1151, 2005.
75. W. Fritsche and M. Hofrichter, "Aerobic degradation by microorganisms," in Environmental Processes- Soil Decontamination, J. Klein, Ed., pp. 146–155, Wiley-VCH, Weinheim, Germany, 2000.
76. R. K. Hommel, "Formation and phylogenetic role of biosurfactants," Journal of Applied Microbiology, vol. 89, no. 1, pp. 158–119, 1990.
77. J. B. Van Beilen and E. G. Funhoff, "Alkane hydroxylases involved in microbial alkane degradation,"Applied Microbiology and Biotechnology, vol. 74, no. 1, pp. 13–21, 2007.
78. T. Zimmer, M. Ohkuma, A. Ohta, M. Takagi, and W.-H. Schunck, "The CYP52 multigene family ofCandida

maltosa encodes functionally diverse n-alkane-inducible cytochromes p450," Biochemical and Biophysical Research Communications, vol. 224, no. 3, pp. 784–789, 1996.

79. U. Scheuer, T. Zimmer, D. Becher, F. Schauer, and W.-H. Schunck, "Oxygenation cascade in conversion of n-alkanes to α,ω-dioic acids catalyzed by cytochrome P450 52A3," Journal of Biological Chemistry, vol. 273, no. 49, pp. 32528–32534, 1998.

80. J. B. Van Beilen and E. G. Funhoff, "Expanding the alkane oxygenase toolbox: new enzymes and applications," Current Opinion in Biotechnology, vol. 16, no. 3, pp. 308–314, 2005.

81. M. O. Ilori, S. A. Adebusoye, and A. C. Ojo, "Isolation and characterization of hydrocarbon-degrading and biosurfactant-producing yeast strains obtained from a polluted lagoon water," World Journal of Microbiology and Biotechnology, vol. 24, no. 11, pp. 2539–2545, 2008.

82. G. S. Kiran, T. A. Hema, R. Gandhimathi, J. Selvin, T. A. Thomas, T. Rajeetha Ravji, and K. Natarajaseenivasan, "Optimization and production of a biosurfactant from the sponge-associated marine fungus Aspergillus ustus MSF3," Colloids and Surfaces B, vol. 73, no. 2, pp. 250–256, 2009.

83. O. S. Obayori, M. O. Ilori, S. A. Adebusoye, G. O. Oyetibo, A. E. Omotayo, and O. O. Amund, "Degradation of hydrocarbons and biosurfactant production by Pseudomonas sp. strain LP1," World Journal of Microbiology and Biotechnology, vol. 25, no. 9, pp. 1615–1623, 2009.

84. M. L. Brusseau, R. M. Miller, Y. Zhang, X. Wang, and G. Y. Bai, "Biosurfactant and cosolvent enhanced remediation of contaminated media," ACS Symposium Series, vol. 594, pp. 82–94, 1995.

85. G. Bai, M. L. Brusseau, and R. M. Miller, "Biosurfactant-enhanced removal of residual hydrocarbon from soil," Journal of Contaminant Hydrology, vol. 25, no. 1-2, pp. 157–170, 1997.

86. T. Barkay, S. Navon-Venezia, E. Z. Ron, and E. Rosenberg, "Enhancement of solubilization and biodegradation of polyaromatic hydrocarbons by the bioemulsifier alasan," Applied and Environmental Microbiology, vol. 65, no. 6, pp. 2697–2702, 1999.
87. S. S. Cameotra and P. Singh, "Bioremediation of oil sludge using crude biosurfactants," International Biodeterioration and Biodegradation, vol. 62, no. 3, pp. 274–280, 2008.
88. R. Beal and W. B. Betts, "Role of rhamnolipid biosurfactants in the uptake and mineralization of hexadecane in Pseudomonas aeruginosa," Journal of Applied Microbiology, vol. 89, no. 1, pp. 158–168, 2000.
89. O. Pornsunthorntawee, P. Wongpanit, S. Chavadej, M. Abe, and R. Rujiravanit, "Structural and physicochemical characterization of crude biosurfactant produced by Pseudomonas aeruginosa SP4 isolated from petroleum—contaminated soil," Bioresource Technology, vol. 99, no. 6, pp. 1589–1595, 2008.
90. M. Nikolopoulou and N. Kalogerakis, "Biostimulation strategies for fresh and chronically polluted marine environments with petroleum hydrocarbons," Journal of Chemical Technology and Biotechnology, vol. 84, no. 6, pp. 802–807, 2009.
91. N. G. Wilson and G. Bradley, "The effect of immobilization on rhamnolipid production byPseudomonas fluorescens," Journal of Applied Bacteriology, vol. 81, no. 5, pp. 525–530, 1996.
92. M. P. Díaz, K. G. Boyd, S. J. W. Grigson, and J. G. Burgess, "Biodegradation of crude oil across a wide range of salinities by an extremely halotolerant bacterial consortium MPD-M, immobilized onto polypropylene fibers," Biotechnology and Bioengineering, vol. 79, no. 2, pp. 145–153, 2002.
93. C. J. Cunningham, I. B. Ivshina, V. I. Lozinsky, M. S. Kuyukina, and J. C. Philp, "Bioremediation of diesel-contaminated soil by microorganisms immobilised in polyvinyl alcohol,"

International Biodeterioration and Biodegradation, vol. 54, no. 2-3, pp. 167–174, 2004.

94. R. N. Z. A. Rahman, F. M. Ghazali, A. B. Salleh, and M. Basri, "Biodegradation of hydrocarbon contamination by immobilized bacterial cells," Journal of Microbiology, vol. 44, no. 3, pp. 354–359, 2006.
95. W. J. Nichols, "The U.S. Environmental Protect Agency: National Oil and Hazardous Substances Pollution Contingency Plan, Subpart J Product Schedule (40 CFR 300.900)," in Proceedings of the International Oil Spill Conference, pp. 1479–1483, American Petroleum Institute, Washington, DC, USA, 2001.
96. R. Z. Hoff, "Bioremediation: an overview of its development and use for oil spill cleanup," Marine Pollution Bulletin, vol. 26, no. 9, pp. 476–481, 1993.
97. U.S. EPA, "Spill NCP Product Schedule," 2002, http://www.epa.gov/oilspill.
98. K. Lee, G. H. Tremblay, J. Gauthier, S. E. Cobanli, and M. Griffin, "Bioaugmentation and biostimulation: a paradox between laboratory and field results," in Proceedings of the International Oil Spill Conference, pp. 697–705, American Petroleum Institute, Washington, DC, USA, 1997.
99. A. Ladousse and B. Tramier, "Results of 12 years of research in spilled oil bioremediation: inipol EAP 22," in Proceedings of the International Oil Spill Conference, pp. 577–581, American Petroleum Institute, Washington, DC, USA, 1991.
100. J. L. Ramos, A. Wasserfallen, K. Rose, and K. N. Timmis, "Redesigning metabolic routes: manipulation of TOL plasmid pathway for catabolism of alkylbenzoates," Science, vol. 235, no. 4788, pp. 593–596, 1987.
101. S. Panke, J. M. Sánchez-Romero, and V. De Lorenzo, "Engineering of quasi-natural Pseudomonas putidastrains toluene metabolism through an ortho-cleavage degradation pathway," Applied and Environmental Microbiology, vol. 64, no. 2, pp. 748–751, 1998.

102. F. Rojo, D. H. Pieper, K.-H. Engesser, H.-J. Knackmuss, and K. N. Timmis, "Assemblage of ortho cleavage route for simultaneous degradation of chloro- and methylaromatics," Science, vol. 238, no. 4832, pp. 1395–1398, 1987.

103. Y. Hrywna, T. V. Tsoi, O. V. Maltseva, J. F. Quensen III, and J. M. Tiedje, "Construction and characterization of two recombinant bacteria that grow on ortho- and para-substituted chlorobiphenyls," Applied and Environmental Microbiology, vol. 65, no. 5, pp. 2163–2169, 1999.

104. B. D. Erickson and F. J. Mondello, "Enhanced biodegradation of polychlorinated biphenyls after site-directed mutagenesis of a biphenyl dioxygenase gene," Applied and Environmental Microbiology, vol. 59, no. 11, pp. 3858–3862, 1993.

105. A. Suyama, R. Iwakiri, N. Kimura, A. Nishi, K. Nakamura, and K. Furukawa, "Engineering hybridpseudomonads capable of utilizing a wide range of aromatic hydrocarbons and of efficient degradation of trichloroethylene," Journal of Bacteriology, vol. 178, no. 14, pp. 4039–4046, 1996.

106. T. C. Zwick, E. A. Foote, A. J. Pollack, et al., "Effects of nutrient addition during bioventing of fuel contaminated soils in an arid environment," in In-Situ and On-Site Bioremediation, pp. 403–409, Battelle Press, Columbus, Ohio, USA, 1997.

107. M. I. Van Dyke, H. Lee, and J. T. Trevors, "Survival of luxAB-marked Alcaligenes eutrophus H850 in PCB-contaminated soil and sediment," Journal of Chemical Technology and Biotechnology, vol. 65, no. 2, pp. 115–122, 1996.

108. B. M. Applegate, S. R. Kehrmeyer, and G. S. Sayler, "A chromosomally based tod-luxCDABE whole-cell reporter for benzene, toluene, ethybenzene, and xylene (BTEX) sensing," Applied and Environmental Microbiology, vol. 64, no. 7, pp. 2730–2735, 1998.

109. G. S. Sayler and S. Ripp, "Field applications of genetically engineered microorganisms for bioremediation processes," Current Opinion in Biotechnology, vol. 11, no. 3, pp. 286–289, 2000.

110. L. Masson, B. E. Tabashnik, A. Mazza, G. Préfontaine, L. Potvin, R. Brousseau, and J.-L. Schwartz, "Mutagenic analysis of a conserved region of domain III in the Cry1ac toxin of Bacillus thuringiensis,"Applied and Environmental Microbiology, vol. 68, no. 1, pp. 194–200, 2002.

111. C. Sousa, V. De Lorenzo, and A. Cebolla, "Modulation of gene expression through chromosomal positioning in Escherichia coli," Microbiology, vol. 143, no. 6, pp. 2071–2078, 1997.

112. C. J. Kelly, C. A. Lajoie, A. C. Layton, and G. S. Sayler, "Bioluminescent reporter bacterium for toxicity monitoring in biological wastewater treatment systems," Water Environment Research, vol. 71, no. 1, pp. 31–35, 1999.

113. A. C. Layton, B. Gregory, T. W. Schultz, and G. S. Sayler, "Validation of genetically engineered bioluminescent surfactant resistant bacteria as toxicity assessment tools," Ecotoxicology and Environmental Safety, vol. 43, no. 2, pp. 222–228, 1999. ·

114. S. Le Floch, F.-X. Merlin, M. Guillerme, C. Dalmazzone, and P. Le Corre, "A field experimentation on bioremediation: bioren," Environmental Technology, vol. 20, no. 8, pp. 897–907, 1999.

115. S. Le Floch, F. X. Merlin, M. Guillerme, et al., "Bioren: recent experiment on oil polluted shoreline in temperate climate," in In-Situ and On-Site Bioremediation, pp. 411–417, Battelle Press, Columbus, Ohio, USA, 1997.

116. J. L. Schnoor, L. A. Licht, S. C. McCutcheon, N. L. Wolfe, and L. H. Carreira, "Phytoremediation of organic and nutrient contaminants," Environmental Science and Technology, vol. 29, no. 7, pp. 318A–323A, 1995.

117. D. Hecht and G. Badiane, "Phytoremediation," New Internationalist, June 1998.

118. K. V. Nedunuri, R. S. Govindaraju, M. K. Banks, A. P. Schwab, and Z. Chen, "Evaluation of phytoremediation for field-scale degradation of total petroleum hydrocarbons," Journal of

Environmental Engineering, vol. 126, no. 6, pp. 483–490, 2000.

119. U.S. Army Corps of Engineers, Agriculturally Based Bioremediation of Petroleum-Contaminated Soils and Shallow Groundwater in Pacific Island Ecosystems, 2003.

120. R. K. Miya and M. K. Firestone, "Enhanced phenanthrene biodegradation in soil by slender oat root exudates and root debris," Journal of Environmental Quality, vol. 30, no. 6, pp. 1911–1918, 2001.

Chapter 5

Experimental Investigation of Fuel Quality and Contaminant Materials from Liquefied Petroleum Fuel

Jae-Kon Kim, Kyong-Il Min, and Eui Soon Yim

Research Institute of Petroleum Technology, Korea Petroleum Quality Distribution & Authority (K-Petro), Cheongju-City, Chungcheongbuk-do, Republic of Korea

ABSTRACT

In this study, the quality characteristics and residue analysis in circulated LPG fuel were investigated experimentally in Korea. Quality characteristics in circulated LPG fuel were examined with samples of LPG in the supply chain (refinery, petrochemical, and imported LPG), transport, gas stations, and vehicles. The experimental results showed that quality of all circulated LPG was well within the quality standard guideline of LPG in Korea. Especially, it has shown average 13 wt ppm in sulfur content over the full circulated LPG. The residue samples in LPG fuel were extracted on 2 L scale with acetonitrile and analyzed by gas chromatography-mass spectrometry (GC-MS). The components of residues in LPG were composed of 62 organic chemicals with C_3 ~ C_{28} and the main ingredients of residue were plasticizers ((di-octyl phalate (DOP), di-octyl adiphate (DOA) etc.), lubricant oil and amine compounds. It was also showed that mass of residue in vehicles was increasing compared with supply (refinery, petrochemical, and imported LPG). It was presumed that this residue had been originated from automotive LPG fuel, vehicle components, and lubricant oil in infrastructure.

INTRODUCTION

Liquefied petroleum gas (LPG) is another alternative fuel with a high octane number. Alternative automotive fuels with lower carbon content than gasoline and diesel fuel, such as LPG with a mixture of propane (C_3H_8) and butane (C_4H_{10}) [1] , have been widely investigated for reducing exhaust emissions [2] -[8]. Because LPG fuel primarily consists of simple hydrocarbon compounds, and the emissions from LPG-driven vehicles contain lower levels of hydrocarbon compounds, nitrogen oxides, sulfur oxides, air toxins and particulates. LPG is a byproduct generated in petroleum refineries, which is widely used as fuels in residential, industrial and vehicle application. The composition of LPG depends on its end use and varies greatly according to season, country, property of the crude oil/gas supply used and refining process.

Usually, LPG contains certain amount of residues such as hydrocarbons with higher vaporization points falling in the range of lubricants oils ($>C_{18}$) [9] . The residues amount in LPG is an important quality control item and it can accumulate in pipes, vaporizers, instruments and regulators, resulting in mal-flow and misreading. Recently, on specific applications, residue concentrations in LPG have to meet industrial codes. For instance, the Australian LPG Association requires the residues concentration below 20 mass ppm for vehicle use [10]. As a low carbon and low polluting fossil fuel, it is recognized by governments around the world for the contribution, and it can make towards improved indoor and outdoor air quality and reduced greenhouse gas emission [11] . Especially, it is recognized as a clean fuel in Korea and a part of the Korean Government energy supply plan. For this reason, while demand for propane is stagnant, demand for butane has increased since 2000. LPG consumption is 8.63 million tonnes in 2011 and nearly 49% of the LPG consumption is for LPG vehicles. In 2011, 2.43 millions of LPG vehicles were registered to be used throughout the nation, with more than 1948 of LPG gas stations in operation in Korea [12] .

The technology for LPG vehicles and LPG fuel in Korea was recognized as a global top. Nevertheless, it recently reported complaints about problems being experienced with the performance of vehicles operating on LPG in Korea. Recently, the problem was being attributed by LPG installer and component suppliers to contaminants in the LPG, an oily substance which was being deposited in vaporizer and injector in vehicles as shown in Figure 1 [13].

Of particular significance was the concern being expressed by a major LPG companies. Under these circumstances, this study was carried out to investigate characteristics of residues in circulated LPG fuel to determine the origin of the offending oily material deposited in the vehicle components, such as vehicle vaporizer and injector in LPG vehicle fuel system.

In this study, the quality characteristics and residue analysis in circulated LPG fuel was investigated experimentally. Quality characteristics in circulated LPG fuel were examined with samples of distribution supply step (refinery, petrochemical, and imported LPG), transport step, gas station step, and vehicle step) in Korea. The residue samples in circulated LPG were extracted with acetonitrile after evaporated on 2 L scale and it was analyzed to determine the origin of residue contaminations by gas chromatography-mass spectrometry (GC-MS) for component analysis.

EXPERIMENTAL

Sample Preparation

The samples of LPG used in the work were obtained from the supply chain as the supply step (refinery, import, petrochemical), transport step, gas station step and vehicle step as shown in Figure 2. The LPG supply system commences at the point where LPG is produced and ends at the point where LPG is dispensed into a vehicle at a gas station. The quality characteristics were tested with 142 samples of LPG in the supply chain for quality specification. The residue component was analyzed with LPG of 2 L to obtain enough residue concentration and in this method sample quantity was increased by 20 times compared with 100 mL as current test method in ASTM D2158-02. The 142 LPG samples for residue analysis were poured in the separatory funnel with 2 L scale and these were evaporated in hood at room temperature for 6 hours. Finally, the residue in separatory funnel was extracted with acetonitrile sometimes and dried to obtain brown oily residue at room temperature as shown in Figure 3.

(a)

(b)

Figure 1: The contamination in vaporizer and injector from LPG vehicles. (a) Vaporizer contamination; (b) Injector contamination.

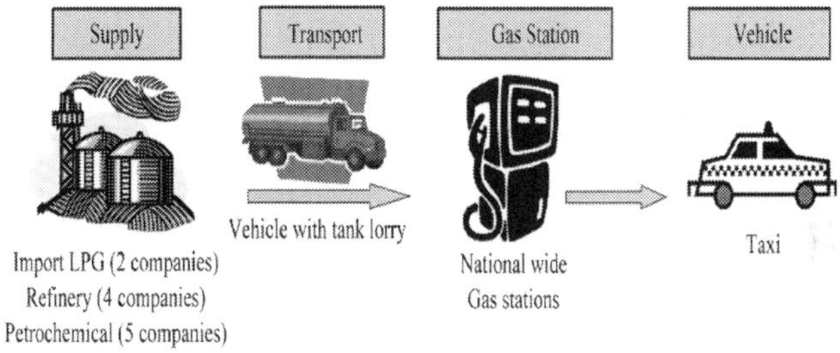

Figure 2: The scheme of experimental sampling of LPG fuel in distribution.

Test Method for LPG Quality

The detailed specification of LPG fuel in this study was listed in Table 1 [14] . The quality of LPG fuel was determined by their physical and fuel properties including composition, sulfur content, vapor pressure, density, residues, and copper corrosion.

LPG Composition

The C3 and C4 composition including olefins in LPG samples were determined by following the KS M ISO 7941, using gas chromatography (GC) (Agilent Technologies 7890A, Palo Alto, USA) equipped with a flame ionization detector (FID) and a chromatography data management system. The fused silica capillary column (SP-2380 model, Supelco, Inc., USA) used in the GC analyzer was 30 m in length and 0.25 mm in inside diameter, with a film thickness of 0.2 μm.

Sulfur Content and Residue Content

For sulfur content, the test was measured by elemental analyzer system NSX-2100V (Mitsubishi Chemical Corporation, Japan) according to ASTM D 6667. The residue was measured by cone-shaped centrifuge tube (Koehler, USA) according to ASTM D 2158 on 100 mL scale.

Vapor Pressure, Density and Copper Corrosion

The vapor pressure was measured by a rapid equilibrium closed cup method (Lawler, USA) according to KS M ISO 4256. The density was measured at 15°C by digital density meter K-25990 (Koehler, USA) using KS M 2150. The copper corrosion was evaluated according to KS M 6151.

All experiments were carried out in triplicate, showing no statistically significant difference between measures. The arithmetic mean of the three determinations was taken as the final result. In addition, the uncertainties were calculated to confirm that each instrument meets the limits of accuracy as set by the specifications of the standard method.

Test Method for LPG Residue Analysis

The condition of GC-MS is showed in Table2 The GC-MS system consisted of an Agilent 7890A GC (Agilent Technologies, Shanghai, China), equipped with an Agilent 7683 auto injector (Agilent Technologies, Shanghai, China), coupled to an Agilent 5975C mass-selective detector (Agilent Technologies, Santa Clara, CA, USA). The GC was fitted with HP-5MS fused silica capillary column (5% phenyl methyl siloxane as non-polar stationary phase, 30 m × 0.25 mm i.d. and 0.25 μm film thickness) from Agilent (J&W Scientific, Folson, CA, USA). The oven temperature program was held at 40°C for 3 min, ramped to 240°C/min and held 240°C for 10 min. Injector and transfer line temperatures were 150°C and 280°C, respectively. The flow rate of the carrier gas (He, 99.9995%) was 1.5 mL/min. A split injection with a ratio of 1:25 was used. The electron impact ionization mass spectrometer was operated as follows: ionization voltage, 70 eV; ion source temperature, 200°C; scan mode, 50.0 - 500.0 (mass range); scan speed, 1.31 scan/sec. The volatile organic compounds in LPG residue were identified by linear retention indices of a series of C_3 to C_{30} by comparison of the mass spectra of each component with the Wiley (Wiley, New York, NY, USA) mass spectral library.

(a)

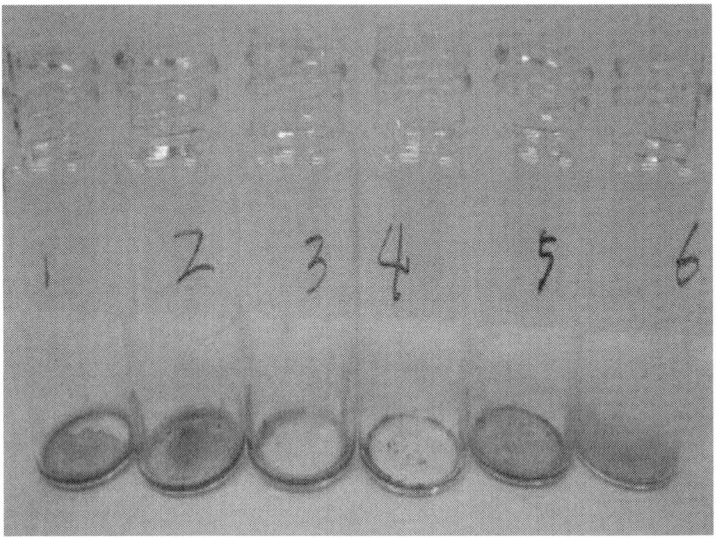

(b)

Figure 3: Extracted residue samples from LPG fuel for GC-MS analysis. (a) Residue in separatory funnel after evaporation; (b) Residue for GC-MS analysis.

Table 1: Test methods for quality specifications of automotive LPG fuel in this experimental

Item		**LPG standard**		**Test method**
Summer		Winter		
Composition (mol %)	C3-Hydrocarbon	Max. 10	15 ~ 35	
	C4-Hydrocarbon	Min. 85	Min. 60	KS M ISO 7941
	1,3-Butadiene	Max. 0.5		
Sulfur content (mg/kg)		Max. 40		ASTM D 6667
Vapor pressure (40°C, kPa)		Max. 1.27		KS M ISO 4256
Density (15°C, kg/m^3)		500 ~ 620		KS M 2150
Residue (ml)		Max. 0.05		ASTM D 2158
Copper strip corrosion (40°C, 1 h)		Min. 1		KS M 6251

Table 2: The conditions of GC-MS for LPG residue analysis

Item	**Condition**
GC-MS model	Agilent 7890A GC/EI/MSD (Agilent 5975 C)
Column	HP 5 MS (5% Phenyl methyl siloxane 30 m, 250 μm)
Detector temperature	280°C
Oven temperature condition	40°C (5 min), 5°C/min, 240°C (10 min)
Carrier gas	He 9 kPa, 1.5 mL/min
Injection amount	1.0 μl

Table 3: Average results for quality characteristics of automotive LPG samples in distribution

Class Item		**LPG standard**		**Distribution**				
Summer		Winter	Supply	Transport	Gas station	Vehicle	Average	
Composition (mol %)	C3-Hydrocarbon	Max. 10	15 ~ 35	1.1	10.3	14.1	6.8	8.7
	C4-Hydrocarbon	Min.85	Min. 60	98.4	91.2	85.3	92.1	91.1
	1,3-Butadiene	Max. 0.5		0.04	-	-	0.06	0.02
Sulfur content (mg/kg)		Max. 40		8.1	14.7	16.4	10.0	12.9
Vapor pressure (40°C, kPa)		Max. 1.27		0.45	0.58	0.57	0.42	0.52
Density (15°C, kg/m^3)		500 ~ 620		578	570	569	574	572
Residue (ml)		Max. 0.05		0.05↓	Max. 0.05	Max. 0.05	Max. 0.05	Max. 0.05
Copper strip corrosion (40°C, 1 h)		1		Max. 1	Max. 1	Max. 1	Max. 1	Max. 1

RESULT AND DISCUSSION

Quality Characteristics of LPG in Distribution

The quality characteristics of LPG fuel were tested with 142 samples in distribution. The experimental results showed that quality of all circulated LPG fuel was well within the limit by Korean specifications as shown in Table 3 Figure 4 shows a general GC chromatogram for composition analysis of butane in LPG fuel. The LPG fuel composition was evaluated 8.7 mol% for C_3 hydrocarbon and 91.1 mol% for C_4 hydrocarbon by GC. Especially, the residue of LPG fuel was within the limit by Korean specifications on 100 mL scale as a current test method (ASTM D 2158) and was showed average 13 wt ppm in sulfur content over the circulated LPG fuel.

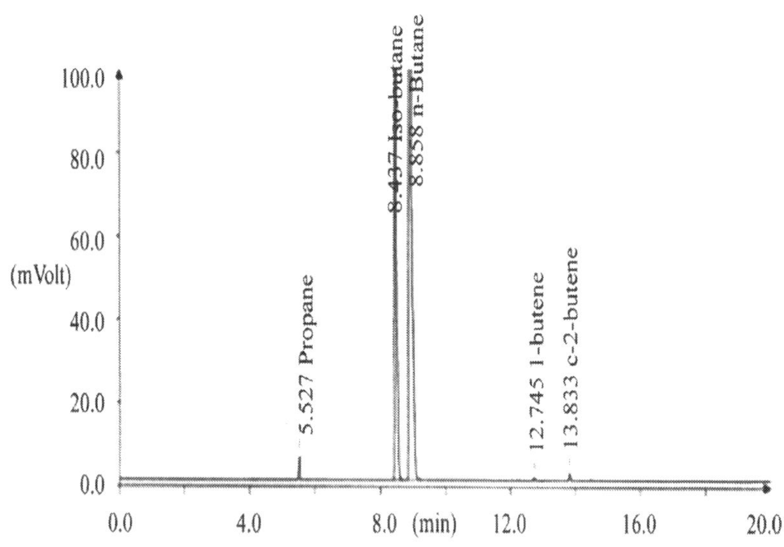

(a)

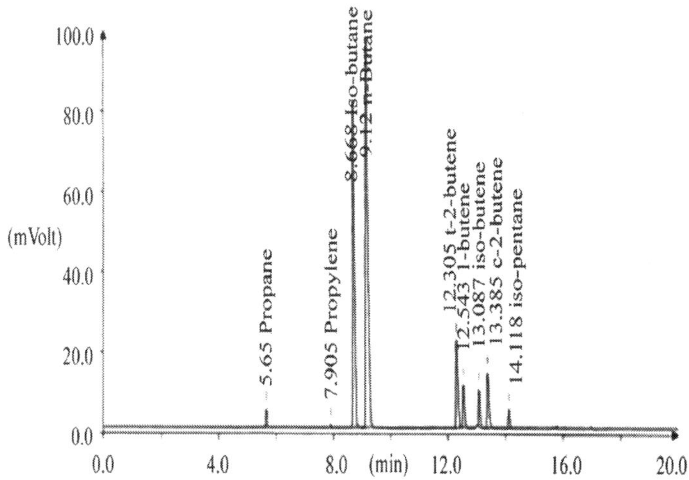

(b)

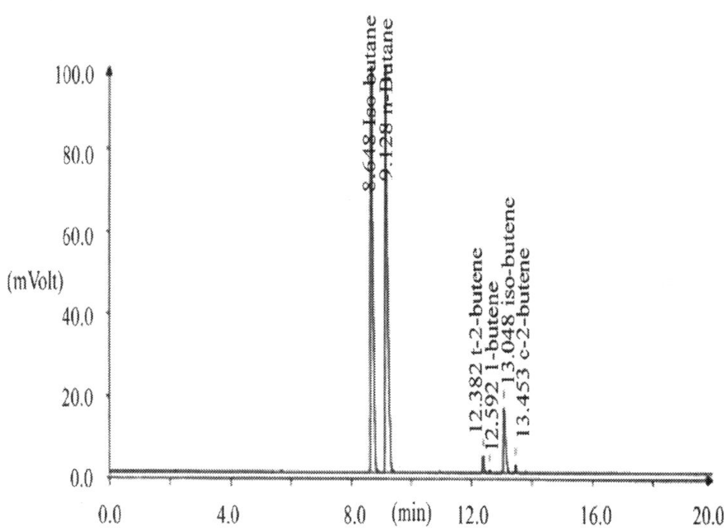

(c)

Figure 4: GC chromatogram for butane composition analysis of LPG fuel in supply step. (a) Importing butane; (b) Refinery butane; (c) Petrochemical butane.

Figure 5 shows a general GC chromatogram for olefin analysis of LPG in vehicle step. The olefin means a sum of unsaturated hydrocarbon such as t-2-butene, 1-butene, iso-butylene, c-2-butene, iso-butene, 1,3-butadiene in LPG composition. Kim et al reported that the olefin content with unsaturated hydrocarbons in LPG has an effect to extract plasticizers [15] [16] . The total olefin composition in distribution also was analyzed to be much as 3.2 mol% average but it was a high level as 7.31 mol% in vehicle step compared with other step as shown in Figure 6. It was statistically increased in the order of transport step < gas station step < vehicle step compared to supply step in distribution.

Residues Analysis of LPG in Distribution

The sources of the residues are the LPG processing equipment, compressors and containers. The amount of residue in LPG is an important quality control item, since it can accumulate in pipes, vaporizers, injectors, instruments and regulators, resulting in mal-flow and misreading. In Korea, the residues content in LPG is measured by ASTM D2158-02 test method. However, it can't give an enough residues concentration to analyze residue component because of a small amount of 100 mL in sampling. To investigate the characteristics of LPG residues for concentration, LPG fuel was evaporated with LPG of 2 L scale to get a more sample compared with contaminants causing problems experienced with the performance of vehicles operating. The residue analysis was tested with 142 samples of LPG in distribution.

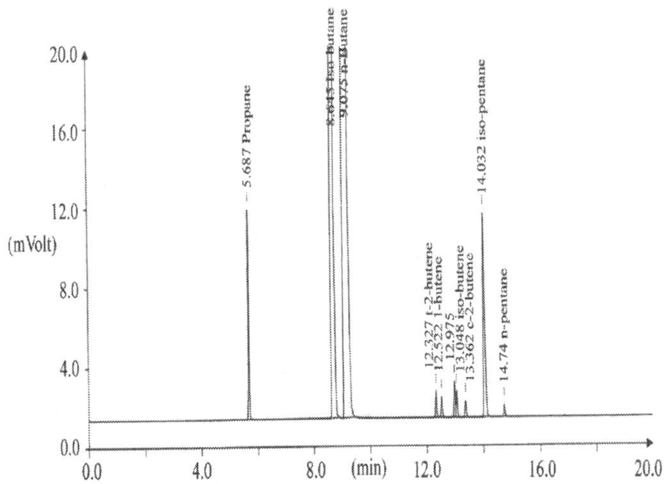

Figure 5: GC chromatogram for butane composition analysis of LPG fuel in vehicle step.

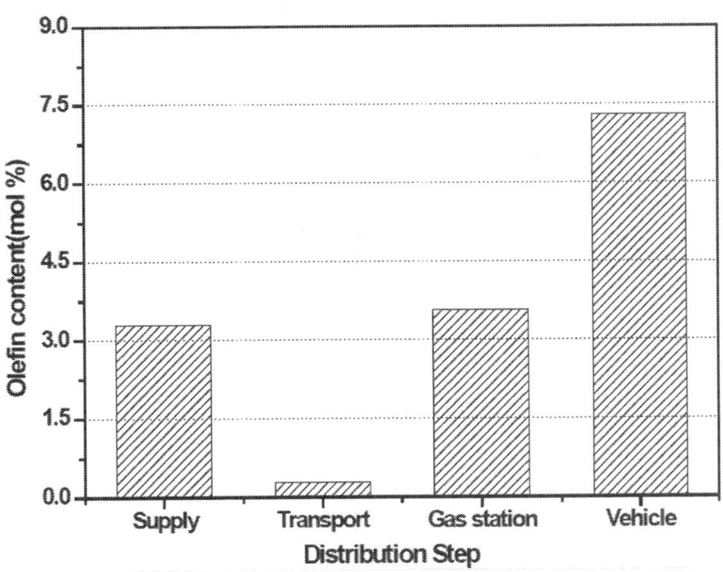

Figure 6: Total olefin content comparison of LPG (142 samples) in distribution (the olefin means a sum of unsaturated hydrocarbon in LPG composition).

The amount of residues comparison of LPG on 2 L scale in distribution is shown in Figure 7. The average total amount of residues in distribution is 19.58 mg/2L and the amount of residue in vehicle step (45.63 mg/2L) is higher than supply step (16.91 mg/2L). It showed that the amount of residue is increasing due to contamination on supply in distribution. Especially, Figure 8 indicated that the residue content of imported LPG was lower than refinery LPG, petrochemical LPG in supply step. The components of residues in LPG were detected62 hydrocarbon chemicals with C_3 ~ C_{28}same as plasticizers, alkane, alkenes and ester with high molecular formula in lubricant oil, amine components, sulfides and siloxane in Table4 Especially, the main ingredients of residue were analyzed plasticizers same as di-n-octylphalate, di-octyladiphate, di(2-ethylhexyl)adipate, di-isooctylalipate, di(isodecyl)phthalate, di-heptyl phthalate by GC-MS in Table5

In general, plasticizers are used in the manufacture of plastics and rubbers to facilitate manufacture and make the products more flexible and softer to meet specific application requirements. Most plasticizers are liquid in their natural state and have an appearance similar to lubricant oil. The harmful plasticizer comprises mostly of either phthalates or adipates or a combination of these chemicals and they were found to have a detrimental effect on the rubber based components. Therefore, LPG residues with high molecular weight have an effect on fuel distribution system in Korea.

CONCLUSIONS

The objective of this research was to assess an investigation of characteristic of residues in circulated LPG fuelto determine the origin of the offending oily material which was being deposited in the vehicle components such as vehicle vaporizer and injector in vehicle fuel system. Recently, the problem was being attributed by LPG installer and component suppliers to contaminants in the LPG an oily substance which deposited in vaporizer and injector in vehicle.

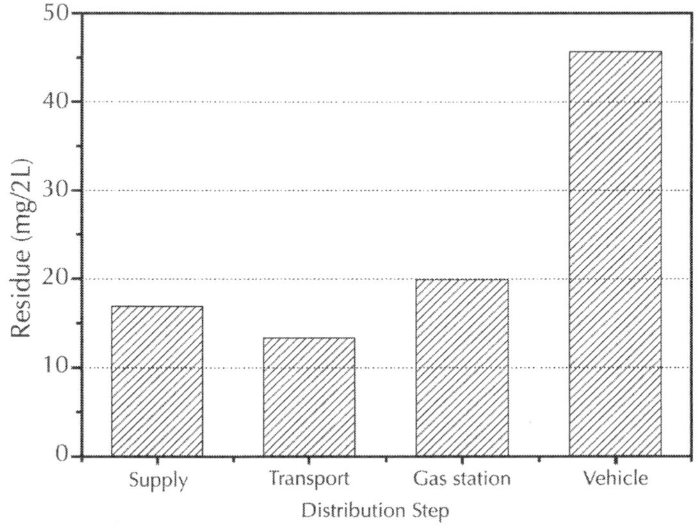

Figure 7: Total amount of residues (mg/2L) comparison of LPG (142 samples) in distribution.

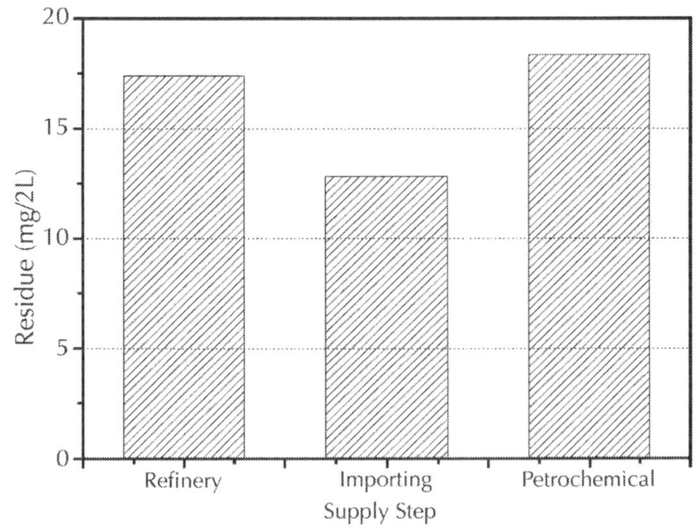

Figure 8: The amount of residues (mg/2L) comparison of LPG (142 samples) in supply step.

Table 4: The residue components from automotive LPG fuel in distribution

Class	Component	Molecular formula	Detection step in distribution[a]
Alkanes	2,2-dimethyl butane	C_6H_{14}	G
	Isodecane	$C_{10}H_{22}$	G
	Undecane	$C_{11}H_{24}$	S
	2,2,6,6-tetramethy-lheptane	$C_{11}H_{24}$	G
	Dodecane	$C_{12}H_{26}$	G
	Tridecane	$C_{13}H_{28}$	G
	2,6,7-trimethyl- decane	$C_{13}H_{28}$	G
	Hexadecane	$C_{16}H_{34}$	G
	Tetracosane	$C_{24}H_{50}$	G
	n-Hexacosane	$C_{26}H_{54}$	G
	n-Octacosane	$C_{28}H_{58}$	S
Amines	Ethamine	C_2H_7N	S
	1,3,4-Thiadiazol-2-amine	$C_2H_3N_3S$	S
	1,2-Ethanediamine	$C_2H_8N_2$	G
	Silanamine (H_5NSi)	H_5NS	G
	N,N-dimethyl formamide	C_3H_7NO	G
	Methyldiethanol amine (MDEA)	$C_5H_{13}NO_2$	G
	2,4,6-Trimehtyl-5H-1,3,5-dithiazin	$C_6H_{13}NS_2$	S
	2-heptanamine	$C_7H_{17}N$	S
	Diphenylamine	$C_{12}H_{11}N$	T
	Octyl-diphenylamine	$C_{20}H_{27}N$	T
Alkens	2,4,6-dodecatriene	$C_{12}H_{20}$	S
	Dimethyl-formyl thiacyclo-hexene	$C_8H_{13}SO$	G
Siloxane	Ttrisiloxane	$H_8O_2Si_3$	S
	Pentasiloxane	$H_{12}O_4Si_5$	G
	Cyclohexasiloxane	$H_{12}O_6Si_6$	S, G

Sulfides	Diethyldisulfide	$C_4H_{10}S_2$	T
	Methyl tert-butyl disulfide	$C_5H_{12}S_2$	T
	Bis(1,1-dimethylethyl) disulfide	$C_8H_{18}S_2$	T
	Dibutyldisulfide	$C_8H_{18}S_2$	S
	Di-n-octyl disulfide	$C_{16}H_{34}S_2$	S
Esters	Methyl cis-3-chloroprope-nate	$C_4H_5ClO_2$	G
	Hexadecanoic acid methyl ester	$C_{17}H_{34}O_2$	G
	Hexadecanoic acid methyl ester	$C_{19}H_{36}O_2$	G
	Hexanedioic acid dioctyl ester	$C_{22}H_{42}O_4$	S

[a]S: supply step, T: transport step, G: gas station step, V: vehicle step

Table 5: The main ingredients of residues contaminations from automotive LPG fuel in distribution

Component	**Molecular formula**	**Detection step in distribution**[a]
Di(2-ethylhexyl)adipate	$C_{22}H_{40}O_4$	S, T, G, V
Di-isooctylalipate	$C_{22}H_{42}O_4$	S, V
Di(isodecyl)phthalate	$C_{28}H_{44}O_4$	G
Di-heptyl phthalate	$C_{22}H_{34}O_4$	G
Bis(n-octyl)phthalate	$C_{24}H_{36}O_4$	G, T
Di(2-ethylhexyl)phthalate	$C_{24}H_{38}O_4$	S, T, G
Di-nonyl phthalate	$C_{26}H_{42}O_4$	G
Di(7-methyloctyl)phthalate	$C_{26}H_{40}O_4$	G

[a]S: Supply step, T: Transport step, G: Gas station step, V: Vehicle step

[a]S: Supply step, T: Transport step, G: Gas station step, V: Vehicle step.

The experimental results showed that quality of all circulated LPG was well within quality standard guideline of LPG in Korea. Especially, it showed average 13 wt ppm in sulfur content over the full circulated LPG. The components of residues in LPG was detected 62 organic chemicals with C_3 ~ C_{28}hydrocarbon compounds and the main ingredients of residue were plasticizers ((di-octylphalate (DOP), di-octyladiphate (DOA) etc.), lubricant oil and amine components. Finally, it was presumed that this residue had been originated from automotive LPG fuel, vehicle components and lubricant oil in infrastructure. LPG residue with high molecular weight has an effect on fuel distribution system in Korea.

ACKNOWLEDGEMENTS

This research was supported by Korea LPG Association (KLPGA) and we sincerely appreciate the support of KLPGA.

REFERENCES

1. Keith, O.K. (1995) Automotive Fuels Reference Book. Society of Automotive Engineers, USA.
2. Boretti, A. (2009) Development of a Direct Injection High Efficiency Liquid Phase LPG Spark Ignition Engine. SAE Paper, No. 2009-01-1881.
3. Chiriac, R., Raud, R., Niculescu, D. and Apostolescu, N. (2003) Development of a LPG Fueled Engine for Heavy Duty Vhicles. SAE Paper, No. 2003-01-3261.
4. Goto, S., Lee, D., Shakal, K., Harayama, N. and Ueno, H. (1999) Performance and Emission of an LPG Lean-Burn Engine for Heavy Duty Vehicles. SAE Paper, No. 1999-01-1513.
5. Kim, C., Lee, D., Oh, S., Kang, K., Chol, H. and Min, K. (2002) Enhancing Performance and Combustion of an LPG MPI Engine for Heavy Duty Vehicles. SAE Paper, No. 2002-01-0449.

6. Lee, J.W., Do, H.S., Kweon, S.I., Park, K.K. and Hong, J.H. (2010) Effect of Various LPG Supply Systems on Exhaust Particle. International Journal of Automotive Technology, 6, 119-124.
7. Myung, C.L., Lee, H., Choi, K., Lee, Y.J. and Park, S. (2009) Effects of Gasoline, Diesel, LPG and Low-Carbon Fuels and Various Certification Modes on Nanoparticle Emission Characteristics in Light-Duty Vehicles. International Journal of Automotive Technology, 10, 537-544. http://dx.doi.org/10.1007/s12239-009-0062-9
8. Park, K. and Han, S. (2004) The Ignition Characteristics for the Direct Liquid Injection Combustion in LPG Engine. FISITA Paper, No. 20042357.
9. Zhuang, Q., Yodatani, J. and Kato, M. (2005) Accurate Measurement Method for the Residues in Liquefied Petroleum Gas (LPG). Fuel, 84, 443-446.http://dx.doi.org/10.1016/j.fuel.2004.09.005
10. Government of Western Australia (2008) Investigation of the Cause and Consequences of Contaminants in Autogas Vehicle Systems. 2008 Report.
11. Goto, S., Wakasa, R. and Lee, D. (2001) Research Trends of LPG-Fueled Engine System. Journal of the Society of Automotive Engineers of Japan, 55, 30-37.
12. Korea LPG Association (2011) Statistical Review of Global LP Gas. 2011-Report.
13. Korea LPG Association (2009) Study of the Origin of Unknown Materials in LPG Distribution. 2009 Report.
14. Korea Petroleum Quality & Distribution Authority (2014) Fuel Quality Standards for Liquefied Petroleum Gas in Korea. http://www.kpetro.or.kr/sub.jsp?MenuID=m2as402
15. Kim, Y.U. (2002) The Component and Compositional Analysis of Trace Materials in LPG. Journal of the Korean Chemical Society, 46, 317-322.http://dx.doi.org/10.5012/jkcs.2002.46.4.317

16. Kim Y.U. (2004) Extraction Property of Plasticizer in LPG High Pressure Rubber Hose. Journal of the Korean Chemical Society, 48, 156-160.http://dx.doi.org/10.5012/jkcs.2004.48.2.156.

Chapter 6

Finite Element Analysis of In-plane Displacements and Von-mises Stresses in Ellipsoidal and Circular Cylinderical Petroleum Tankers

Oluleke Oluwole and Eyere Emagbetere

Mechanical Engineering Department, University of Ibadan, Ibadan, Nigeria

ABSTRACT

Road tankers are the most used means of transporting petroleum product to end users due to its cost effectiveness and energy-efficiency. The cylindrical tank has been well designed for by

ASME VIII divisions 1 and 2 using analytical equations. Petrol tankers are not circular but elliptical probably for stability during transportation. This paper has used the finite element method to investigate in-plane displacements and Von-Mises stresses in both circular and elliptical cylindrical tanks under full loading. An elliptical OANDO® tanker of 66.78 m^3 volume and shell thickness of 0.2 mm and an equivalent volume circular cylindrical tank was used for the simulation. MATLAB® was used to generate geometrical mesh model of the petroleum tankers, extract element coordinates and conduct the finite element analysis. Plane strain condition was used in analyzing a section of the petroleum tanker. It was observed that an equivalent volume circular cylindrical tank was under a higher internal pressure (16,858 N/m^2) compared to the elliptical cylinder (14,480 N/m^2). Von-Mises stress and in-plane displacements showed direct linear relationships with internal fluid pressure. Von-Mises stress in the elliptical tank was found to be lower (5.7×10^6 N/m^2) than for the circular tank (8×10^6 N/m^2). In plane displacements was zero in the longitudinal direction for both tanks and of the order of 10^{-4} mm in the y-direction for both tanks with the circular larger by about 2.5×10^{-3} cm. So in addition to tank stability on the lorry, the Von-Mises stresses were lower as well for the elliptical tank. It was also observed that Von-Mises stresses were far below the yield stress of the steel plate. However, the effect of weldment area on lowering of yield stress was not studied. Stress values were validated using analytical method and found to be insignificantly different ($P > 0.05$).

INTRODUCTION

Road tankers are cylindrical or ellipsoidal pressure vessels used to convey liquids especially petroleum products. Petroleum tanker is widely used over other means of transporting petroleum products (pipeline, trains, badges and trucks) due to its cost effectiveness and energy-efficiency [1]. According to the projection of Ekpen [2], Nigeria's petroleum product consumption will rise to about 300 Kb/day, a large part ofher total energy consumption in the country.

Recently, there has been increasing number of reports of tanker explosion during transit, claiming several lives and properties [3-9]. There is therefore need to consider petroleum tanker safety in line with design.

Road tankers constructed locallyare designed based on ASME standards for pressure vessels (ASME VIII divisions 1 and 2). The ASME standard presents design formulae that are simple to use, but limited to specific geometries and geometry details such as welded supports and openings. It does not put into consideration several actions or combination of actions such as local loads, seismic load, wind loads and external pressure in its design formula. A better approach is the design by analysis. According to Josef et al. [10], design by analysis is very flexible, it allows for any combination of actions, it can be used for complicated geometry, and it addresses directly the creativity of the designer. Finite element method is an efficient technique of design by analysis as it gives result close to corresponding experimental values.

The Finite Element Method is a numerical technique ideally suited to digital computers in which a continuous elastic structure (continuum) is divided (discretized) into smaller but finite well defined sub-structures (element) that can be represented by simple equations [11]. According to George [12], it represents an approximate numerical solution of a boundary value problem described by a differential equation.

There are several publications on application of Finite element analysis of specific pressure vessels. Mirko et al., [13] applied Finite element method in the determination of the stress state of a cylindrical vessel with variable thickness. The stress distribution of deformed tank i.e. a tank with part that has been formally repaired was analyzed [14]. Jaroslav [15], In his work made a listings of over 865 papers and conference proceedings on finite element in the analysis of pressure vessels and pipelines which he classified into the following categories: linear and nonlinear, static and dynamic; stress and deflection analyses; stability problems; thermal problems; fracture mechanics problems; contact problems; fluid-structure interaction problems; manufacturing of pipes and tubes; welded pipes and pressure vessel components; development of special finite elements for pressure vessels and pipes; finite element software.

Most Finite element analysts develop own codes using MATLAB, C++, or FORTRAN. Meshing of geometrical models is done manually giving poor results. Other challenges faced by finite element analysts include: high cost of acquiring CAD software for numerical simulation, difficulty in managing read data from standard CAD programs such as dwg format (AutoCAD file) when using own code, and erroneous results, due to improper use of commercial software.

MATLAB is a computer language and the software has an interactive computing environment that enables numerical computation, analysis and data visualization [16]. This work has explored MATLAB's plotting capability and powerful computational tools to generate the geometrical mesh model of a petroleum tanker, extract element coordinates and conduct a finite element analysis of Von-Mises stress on a loaded tanker. The values obtained were compared with ASME standard.

METHODOLOGY

Automated Mesh Generator

The ellipsoidal tanker which is predominantly more common compared to the cylindrical ones has three inner segments (Figures 1 and 2). Each segment can be divided into four symmetrically similar sections (Figure 3). For ease and simplicity, the left bottom section was considered (Figure 4), and an automated mesh generation code was written with MATLAB®.

The common ellipse (Figure 5) and the ellipse Equation (1) was used in the development of the code. The automated mesh generator was developed using piecewise straight lines to approximate the curve and element continuity created during the meshing relations. The flowchart for the automated mesh generator is shown (Figure 6). Figure 7 shows meshing of the quarter section.

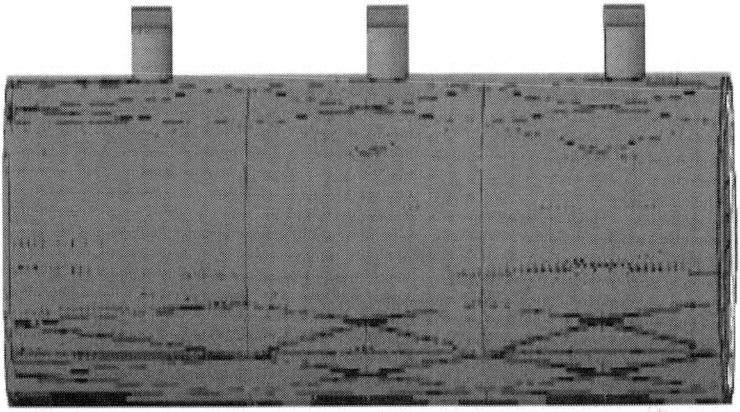

Figure 1: The tank model.

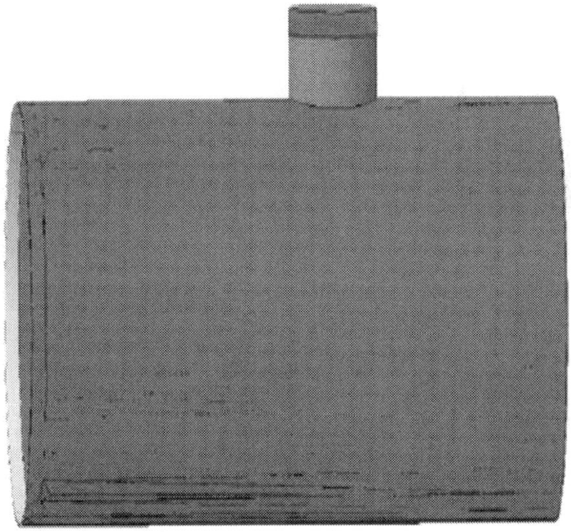

(a)

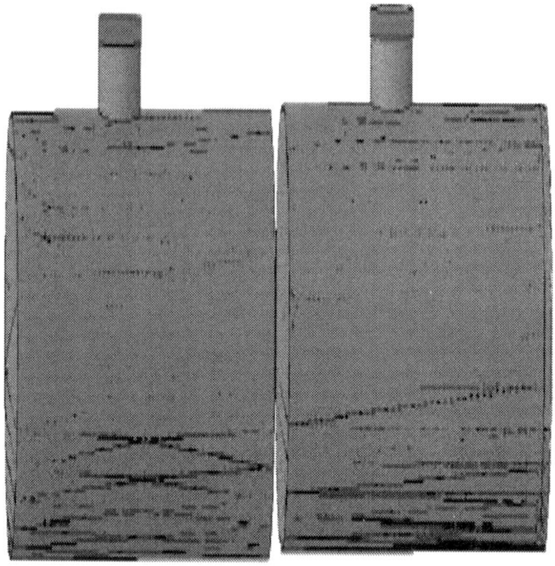

(b)

Figure 2: The tanker model divided into 3 parts initially joined together.

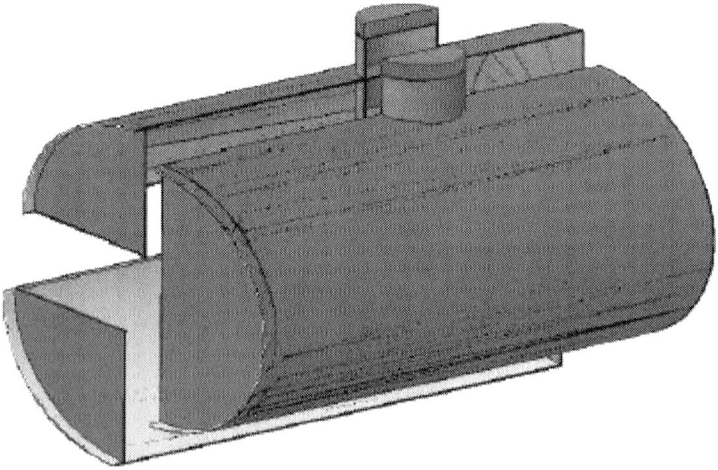

(a)

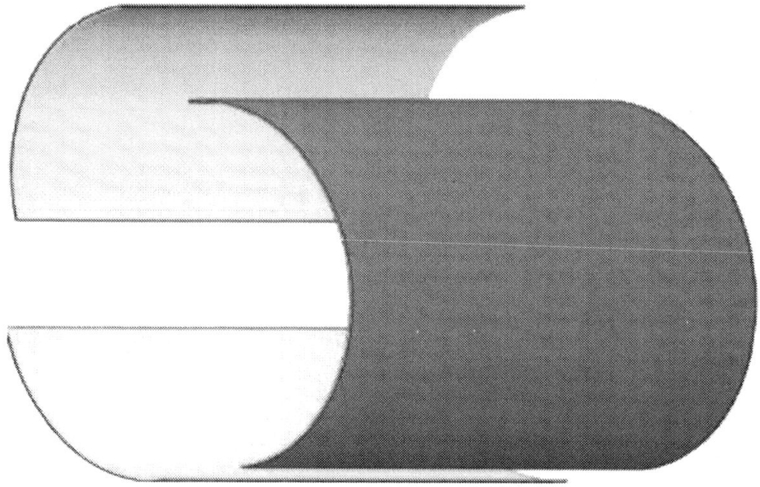

(b)

Figure 3: A part of the tanker further sectioned into parts.

Figure 4: Section for automated meshing.

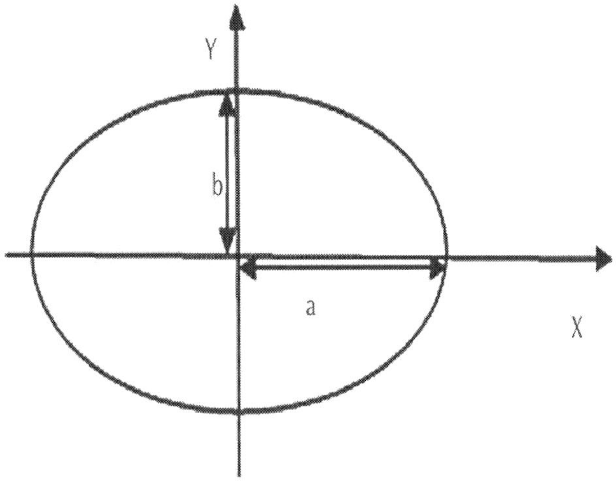

Figure 5: An ellipse with its dimension.

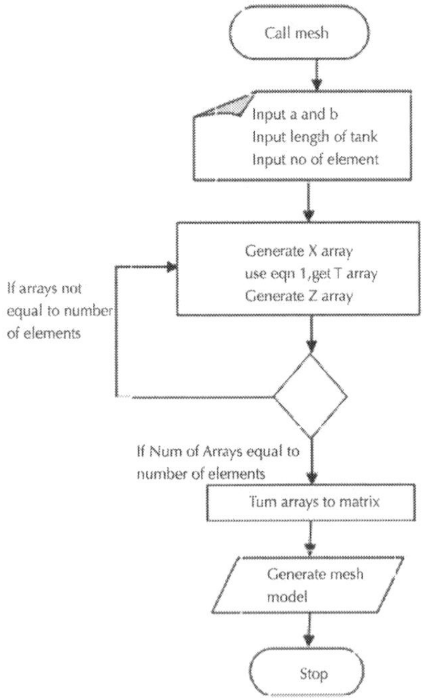

Figure 6: Algorithm for automatic mesh generator.

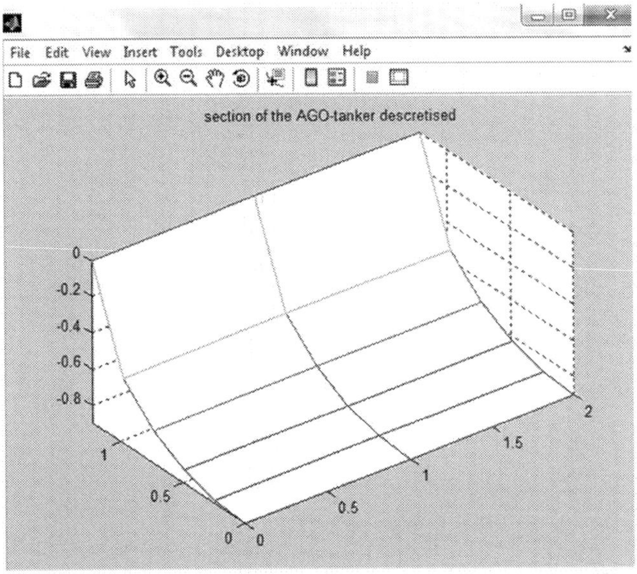
File Edit View Insert Tools Desktop Window Help
section of the AGO-tanker descretised
0
-0.2
-0.4
-0.6
-0.8
1
0.5
0
0
0.5
1
1.5
2

(a)

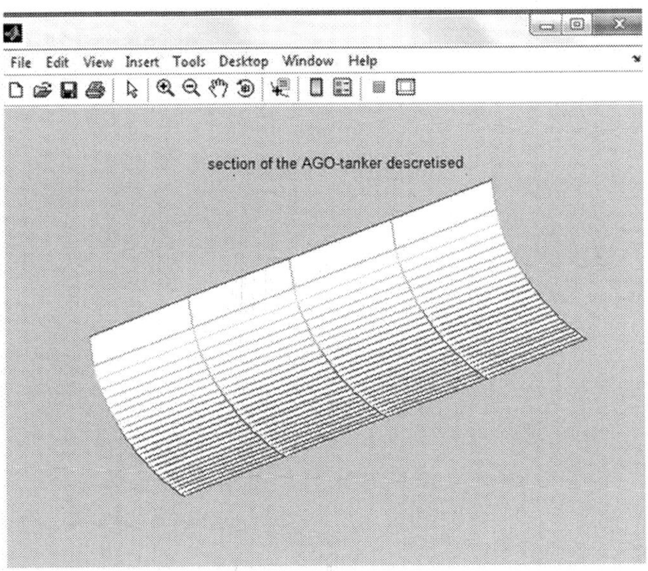
File Edit View Insert Tools Desktop Window Help
section of the AGO-tanker descretised

(b)

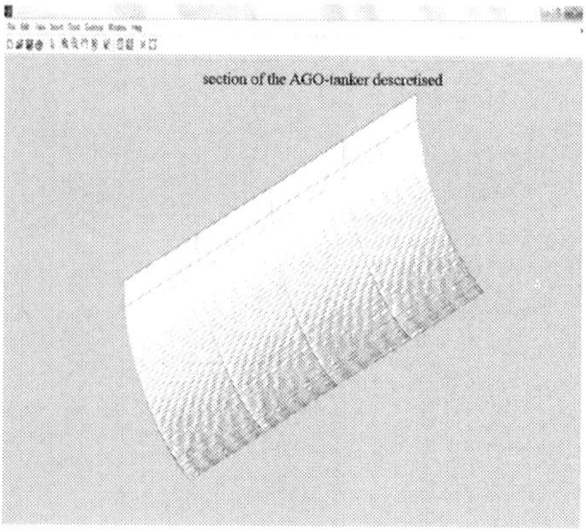

(c)

Figure 7: (a) 10 elements mesh; (b) 200 elements mesh; (c) 500 elements mesh.

$$\frac{\mathbf{X}^2}{\mathbf{a}^2}+\frac{\mathbf{Y}^2}{\mathbf{b}^2}=1$$ also was written as

$$\mathbf{Y}=\sqrt{\mathbf{b}^2-\frac{\mathbf{b}^2\times\mathbf{X}^2}{\mathbf{a}^2}} \qquad (1)$$

where a represents half the horizontal axis, and b half the vertical axis as shown in Figure 5.

Estimation of Stiffness Matrix and Force Vector

After generating the meshed model, an element was extrapolated and used in developing the stiffness matrix and force vector, needed to evaluate the displacement vector for plane strain elasticity of the model and VonMises stress.

The equation for the rectangular shell elements was derived by combining the linear rectangular element (membrane element) and the plate element (bending element).

Membrane Element

The equilibrium equations for a three-dimensional linear elasticity express in terms of stress is given [17,18] as;

$$\sigma_{ij,j} + f_i = 0$$

where $s_{ij,j}$ are the stress components, f_i are the body forces.

The above equilibrium equation expressed in terms of displacement for two dimensional analyses is given as follows. The equation was reduced to two dimensional since plane strain condition is being considered.

$$-\frac{\partial}{\partial x}\left(A11\frac{\partial h_x}{dx} + A12\frac{\partial h_y}{\partial y}\right) - \frac{\partial}{\partial y}\left[A66\left(\frac{\partial h_x}{\partial y} + \frac{\partial h_y}{\partial x}\right)\right] = f_x$$

$$-\frac{\partial}{\partial x}\left(A21\frac{\partial h_x}{dx} + A22\frac{\partial h_y}{\partial y}\right) - \frac{\partial}{\partial y}\left[A66\left(\frac{\partial h_x}{\partial y} + \frac{\partial h_y}{\partial x}\right)\right] = f_y$$

h_x and h_y are the in-plane displacement in the x and y directions respectively of the Cartesian coordinate system.

Weak Formulation of the Membrane Element Equation

$$0 = \int_{y1}^{y2}\int_{x1}^{x2} t\left[\frac{\partial w_i}{\partial x}\left(A11\frac{\partial h_x}{\partial x} + A12\frac{\partial h_y}{\partial y}\right)\right.$$

$$\left. + A66\frac{\partial w_1}{\partial y}\left(\frac{\partial h_x}{\partial y} + \frac{\partial h_y}{\partial y}\right) + \rho w_{2h_x}\right]\mathrm{d}x\mathrm{d}y$$

$$-\int_{y1}^{y2}\int_{x1}^{x2} tw_i f_x \mathrm{d}x\mathrm{d}y - \int_{re} tw_i t_x \mathrm{d}s$$

$$0=\int_{y1}^{y2}\int_{x1}^{x2} t\left[\frac{\partial w_i}{\partial y}\left(A21\frac{\partial h_x}{\partial x}+A22\frac{\partial h_y}{\partial y}\right)\right.$$

$$\left.+A66\frac{\partial w_i}{\partial x}\left(\frac{\partial h_x}{\partial y}+\frac{\partial h_y}{\partial y}\right)+\rho w_{ih_y}\right]\mathrm{d}x\mathrm{d}y$$

$$-\int_{y1}^{y2}\int_{x1}^{x2} tw_i f_y \mathrm{d}x\mathrm{d}y-\int_{re} tw_i t_y \mathrm{d}s$$

Stiffness Matrix for the Membrane Element

The finite element equation is expressed as

$$[K]\{U\}=\{F\}$$

[K] is the stiffness matrix, {F} is the force vector and {U} is the displacement vector.

For plane strain condition, assuming homogeneity of material, E and v are the material elastic constant and poison ratio respectively.

$$A11=A22=\frac{E(1-v)}{(1+v)(1-2v)};$$

$$A12=A21=vA11;\ A66=\frac{E}{2(1+v)}$$

Interpolating Function for the Membrane Element

h_i=[w_i]{h} where {h} is the in-plane displacement vector, [w_i] are the weight functions. The weight function for the rectangular element was used in the analysis.

Bending Element

The standard equation governing plate bending [19] is given as follows;

$$\frac{\partial^2}{\partial x^2}\left(B_{11}\frac{\partial^2\omega}{\partial x^2}+B_{12}\frac{\partial^2\omega}{\partial y^2}\right)+\frac{\partial^2}{\partial y^2}\left(B_{12}\frac{\partial^2\omega}{\partial x^2}+B_{22}\frac{\partial^2\omega}{\partial y^2}\right)$$

$$+2\frac{\partial^2}{\partial x\partial y}\left(2B_{66}\frac{\partial^2\omega}{\partial x\partial y}\right)=q$$

B_{ij} is the bending stiffness matrix for the shell element. E, v and t are the elastic modulus, poison ratio and thickness respectively of the tanker being considered.

$$B_{11}=\frac{Et^3}{12\left(1-v^2\right)};B_{22}=\frac{Et^3}{12\left(1-v^2\right)};$$

$$B_{12}=\frac{vEt^3}{12\left(1-v^2\right)};B_{66}=\frac{Gt^3}{12}$$

Weak Formulation of the Bending Element

$$0=\int_\Omega\left[\frac{\partial^2 U}{\partial x^2}\left(B_{11}\frac{\partial^2\omega}{\partial x^2}+B_{12}\frac{\partial^2\omega}{\partial y^2}\right)\right.$$

$$+\frac{\partial^2 U}{\partial y^2}\left(B_{12}\frac{\partial^2\omega}{\partial x^2}+B_{22}\frac{\partial^2\omega}{\partial y^2}\right)$$

$$\left.+4B_{66}\frac{\partial^2 U}{\partial x\partial y}\frac{\partial^2\omega}{\partial x\partial y}-I_0 v\right]\mathrm{d}x\mathrm{d}y$$

$$-\int_v\left[\left(\frac{\partial M_{xx}}{\partial x}+\frac{\partial M_{xy}}{\partial y}\right)n_x+\left(\frac{\partial M_{xy}}{\partial x}+\frac{\partial M_{yy}}{\partial y}\right)n_y\right]\mathrm{d}s$$

$$+\int_v\left[\frac{\partial U}{\partial x}\left(M_{xx}n_x+M_{xy}n_y\right)+\frac{\partial U}{\partial y}\left(M_{xy}n_x+M_{yy}n_y\right)\right]\mathrm{d}s$$

Stiffness Matrix for the Bending Element

For the plate element, on application of the classical plate theory [19], the Finite Element Model is given as follows

$$\left[K^e\right]\left\{\nabla^e\right\}=\left\{F^e\right\}+\left\{Q^e\right\}$$

Every term retains their meaning as above except for ∇ which is used to represent one out-of-plane displacement in the thickness axis and two rotations. Mass matrix was neglected since dynamics of the system was not considered.

Interpolation Function for the Bending Element

$\omega_i = [U_i]\{\omega\}$. Where $\{\omega\}$ represents one out-of-plane displacement in the thickness direction and two rotations and $[U_i]$ are the weight functions. B_{ij}' are the material elasticity matrices for the bending element.

$$B_{11}=\frac{E_1 t^3}{12\left(1-v^2\right)}, B_{22}=\frac{E_2 t^3}{12\left(1-v^2\right)}$$

$$B_{12}=\frac{v_{12}E_2 t^3}{12\left(1-v^2\right)}, B_{66}=\frac{G_{12t^3}}{12}$$

Calculation of p and q for Each Element

The stiffness matrices, and force vectors are basically a function of the side length of the rectangular element (Figures 8(a) and (b)) used to form the tanker model. From Figure 8, the values of p and q were calculated applying coordinate geometry as follows:

$$p = \sqrt{(y_4 - y_1)^2 + (x - x_1)^2}$$

$$q = z_2 - z_1$$

x_i, y_i and z_i are the coordinate locations of each element. I = 1 to 4 represents the four nodes of a 4 nodal rectangular element.

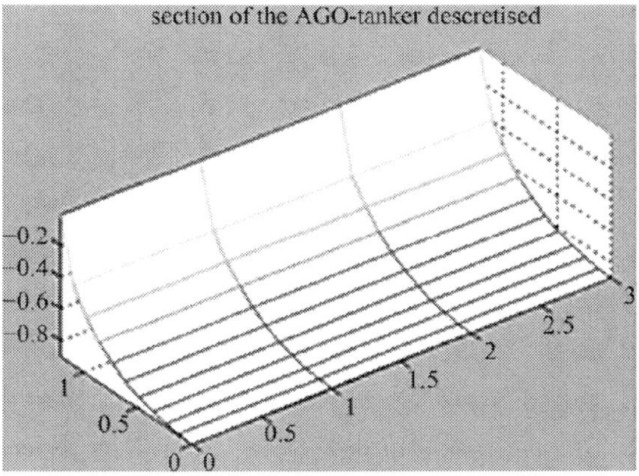

(a)

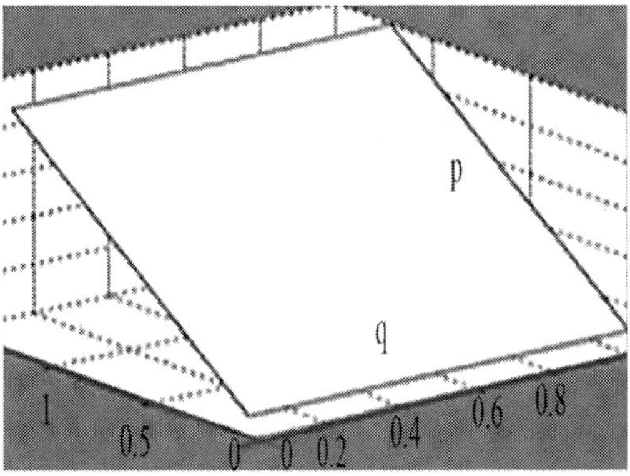

(b)

Figure 8: (a) Element mesh of the section; (b) An element showing dimensions for FEA.

Combination of Membrane and Bending Stiffness Matrices and Force Vectors

$$[K]+[Kb]=[KG]$$

where [K] is the 8 × 8 membrane stiffness matrix and [Kb] is the 12 × 12 bending stiffness matrix, both combining to give the total elemental stiffness matrix [KG] in the form shown below

$$[KG]=\begin{bmatrix} K & 0 \\ 0 & Kb \end{bmatrix}$$

The force vectors were combined in similar manner to form the total elemental force vector as follows

$$[FG]=\begin{Bmatrix} F \\ Fb \end{Bmatrix}$$

Boundary Conditions and Calculation of Displacement

Each element has five degrees of freedom (two in-plane displacements u and v, one out-of-plane displacements w, and two rotations. The displacements and bending at the edges were given zero since they are fixed.

Code for evaluating the finite element equation {U} =[k]/[f], introducing boundary conditions and estimating the displacement was done using the flow algorithm (Figure 9).

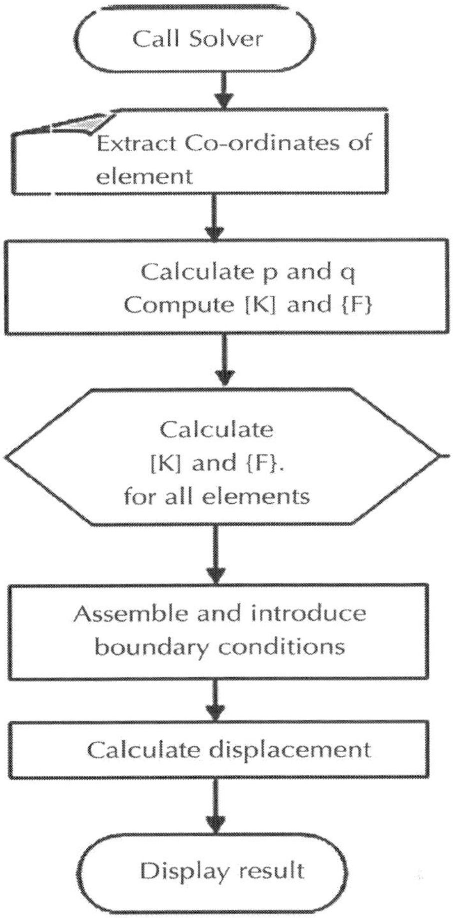

Figure 9: Algorithm applying extracted elemental coordinate for FE analysis.

Calculation of Stresses and Von-Mises Stress

$$[\text{Stress}] = [A] \times [B] \times \{U\}$$

A_{ij} represent the elasticity matrices for both the membrane element and bending element respectively.

[B] Represents the strain vectors and moment vector for membrane and bending element respectively.

[Stress] is the bending and membrane stress being computed.

Analytical Von-mises Stress Calculation to compute the Von-mises stress analytically, there are three principal stresses which are all functions of the pressure.

P1 = Principal stress 1 = Hoop stress = pressure × horizontal radii axis/t;

P2 = Principal stress 2 = Hoop stress/2;

P3 = Principal stress 3 = radial stress pressure;

$$\text{Von-Mises1} = \left(\sqrt{(P1-P2)^2+(P1-P3)^2+(P2-P3)^2}\right)/2$$

Dimensions Used for Analysis

Length of tanker = 485 cm;

Vertical axis of tanker = 180 cm;

Horizontal axis of tanker = 244 cm;

Thickness of tanker = 0.2 cm;

Poison ratio = 0.3;

Material of construction = A516M Grade 70;

Specified minimum yield stress = 25×10^7 N/m^2;

Maximum allowable stress = 13.8×10^7 N/m^2;

Elastic modulus = 200×10^9 N/m^2.

RESULTS AND DISCUSSIONS

Results

Figure 10 shows Von-Moses stress variation with loading. At full loading, Von-Mises stress was 5.7×10^6 N/m^2 and this decreased linearly to zero as the tank is offloaded until there is an empty tank with zero VonMises stress.

Figure 11 shows Von-Mises stress converge to a consistent value at large mesh elements for both elliptical and circular tanks. It also shows Von-Mises stress for circular tanks slightly higher than for the elliptical tank. Both FEM computed values converged to the analytical values for both circular and elliptical. ASME values tallied with the values for circular tank which are the same as the analytical values for the circular tank.

Figure 12 shows the effect of tanker internal pressure on in-plane displacement in the longitudinal (x) direction of elliptical cylindrical tankers. It shows a linear relationship between both parameters consistent with Hooke's law at elastic dimensions.

Figure 13 shows the effect of variation in internal pressure in elliptical tank on in-plane circumferential (y) displacements. It shows as well a linear relationship between both parameters.

Figure 14 shows the combined plots of in-plane longitudinal displacements for both circular and elliptical cylindrical tanks at full loading. The plots show that longitudinal displacements were the same for both tanks and were as it were non-existent (zero).

Figure 15 shows the combined plot of the circumferential in-plane displacements for both tanks at full loading. The plots show that the circumferential in-plane displacements were existent but at an order of $10^{-4\ cm}$. The circular cylindrical tank showed higher circumferential displacement than the elliptical with a difference of about $2.5 \times 10^{-3\ cm}$.

Figure 16 shows Von-Mises stress variation with tanker shell thickness. It shows an increasing stress as shell thickness is reduced and after 5.0 × 10^{-2} m shell thickness, Von-Mises stress increased rapidly till yield stress is reached.

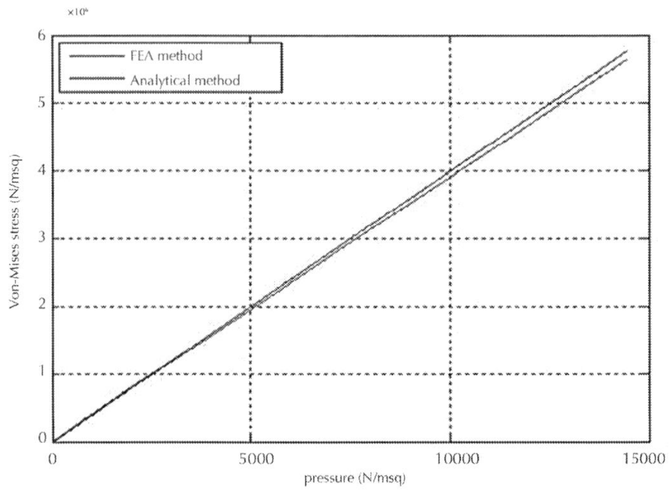

Figure 10: Von-Mises stress variation with internal tank pressure for elliptical cylindrical tank using 800 element meshes.

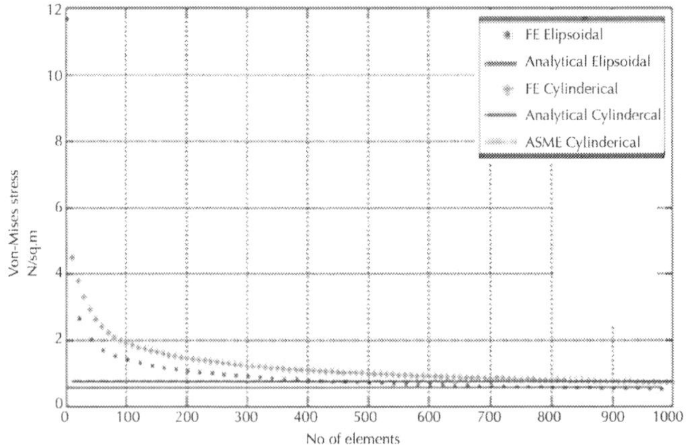

Figure 11: Von-Mises stress variation with no of meshing elements for cylindrical and elliptical cylindrical tank at 16,858 N/m^2 and 14,480 N/m^2loading pressure respectively.

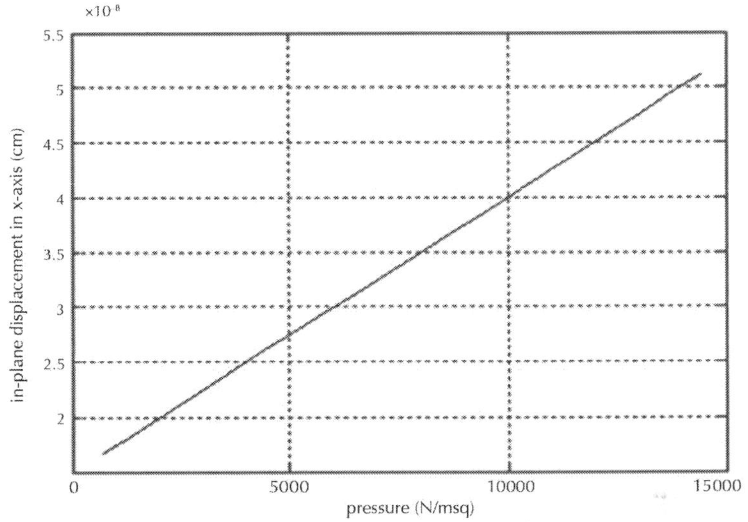

Figure 12: Variation of in-plane displacement in the longitudinal (x) direction of elliptical cylindrical tankers with internal load pressure.

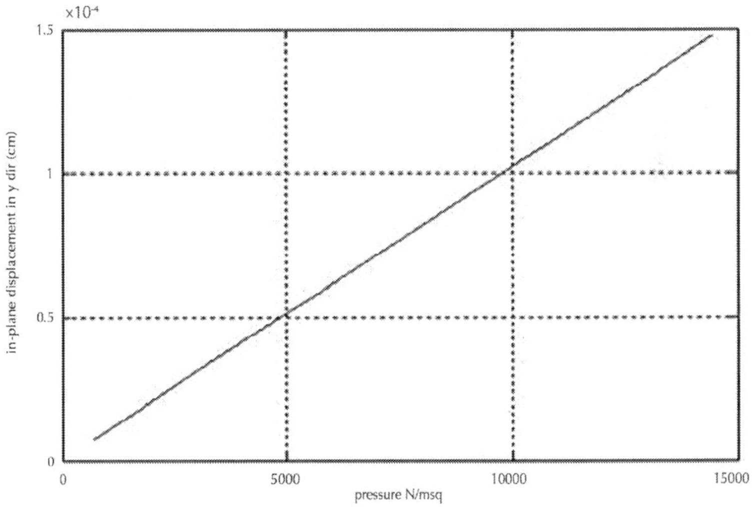

Figure 13: Variation of in-plane displacement in the circumferential (y) direction of Elliptical cylindrical tankers with internal load pressure.

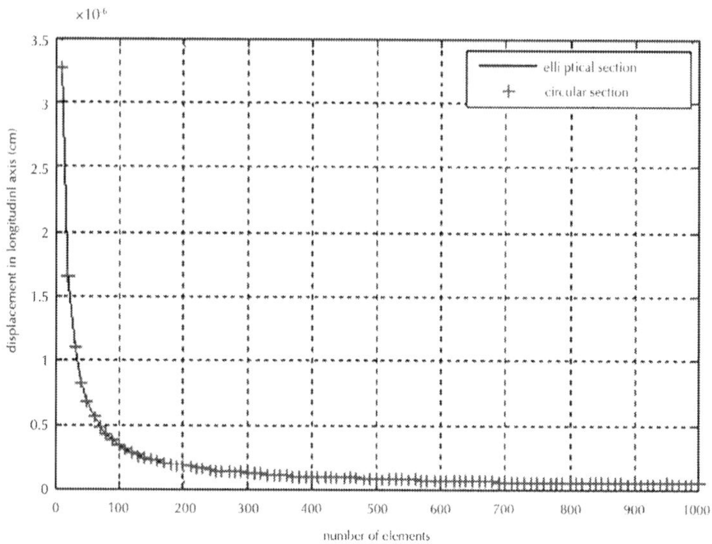

Figure 14: In plane displacement in the longitudinal (x) directionof elliptical and circular cylindrical tankers at full loading.

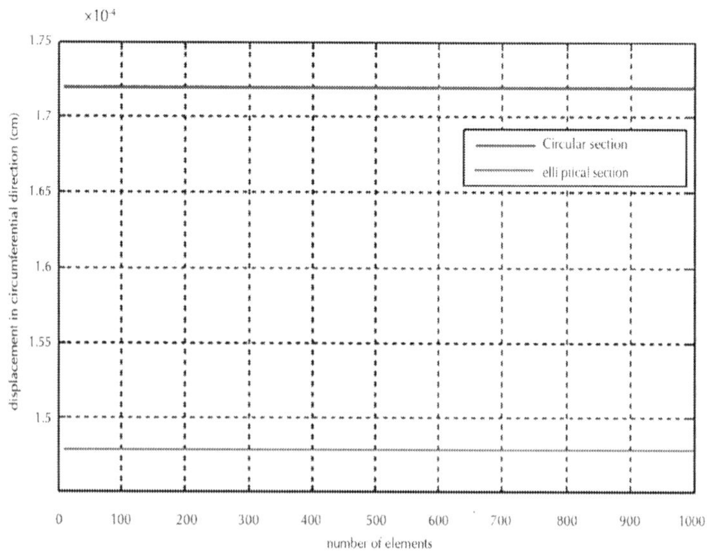

Figure 15: In plane displacement in the circumferential (y) direction of elliptical and circular cylindrical tankers at full loading.

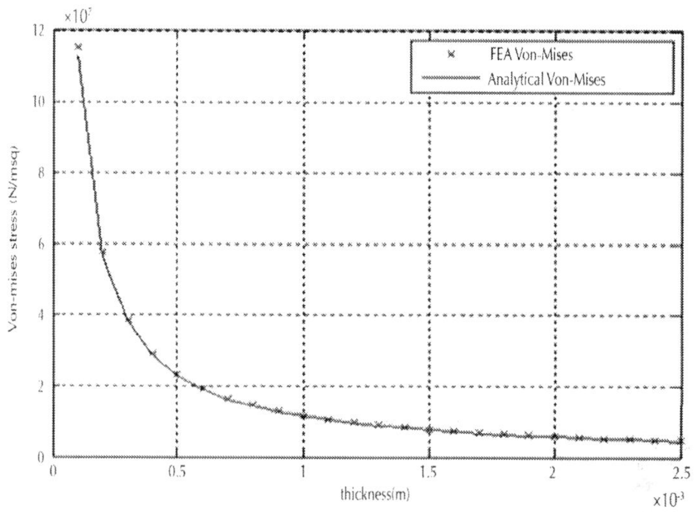

Figure 16: Von-Mises variation with tanker shell thickness.

DISCUSSION OF RESULTS

The direct linear relationships between internal loading pressure and Von-Mises stress (Figure 10) and that between in-plane displacements and internal pressure (Figures 12 and 13) are not unexpected because of the linearity of the stress-strain relationship in the elastic region. The result clearly showed that at full load Von-Mises stress is maximum (Figures 10 and 11). However, the Von-Mises stress for the fully loaded tanker still fell far below the yield stress of the material and the maximum allowable stress (Figure 11). This showed that the tanker will not yield during transportation from the loading pressure. However, since the tanks are always welded and not seamless, the tanks are susceptible to yield stress lowering effect of heat-affected-zone (HAZ) from welding. The HAZ causes embrittlement, hence a lowering of yield stress and if care is not taken, there could be microhole defects from where rapid crack propagation could develop causing catastrophic failure of the tank. If this effect is coupled to the corrosive effect of the petroleum product over time causing what can be termed stresscorrosion effect on the welded portion could as well cause failure over time. However, these issues were not addressed in this paper and are concerns for future work.

The results also show that circular cylinderical tanks have higher internal pressures as well as higher VonMises stresses than elliptical tanks of the same loading volume (Figure 13). The lower Von-Mises stresses developed in elliptical cylindrical tanks as well as their stability during transportation gives the elliptical tank a big edge over the circular cylindrical tank.

The in-plane displacements (Figures 14 and 15) showed good tanker material integrity and nullifies any idea of failure by yielding but just as had been mentioned earlier future work on weldment defects coupled with stresscorrosion need to be done to investigate what is transpiring here which could lead to other types of failure over time.

The inverse and non-linear relationship between tank shell thickness and Von-Mises stress (Figure 16) was also not surprising because of the relations between hoop/longitudinal stresses and shell thickness. This result showed clearly that it is important to design for a thickness that would guarantee a sufficient factor of safety especially since the effect of the welded parts' heat affected zone (HAZ) on lowering of yield stress must be considered.

CONCLUSIONS

Von-Mises stresses in circular and elliptical petroleum tankers under full loading have been obtained using the finite element method. It was observed that an equivalent volume circular cylindrical has a higher internal pressure (16,858 N/m^2) compared to the elliptical cylinder (14,480 N/m^2). Von-Mises stresses showed a linear relationship with variation in internal fluid pressure. Von-Mises stresses in the elliptical tank were found to be lower (5.7×10^6 N/m^2) than for the circular tank (8×10^6 N/m^2. VonMises stresses were far below the yield stress of the steel plate (25×10^7 N/m^2). However, the effect of weldment area and stress corrosion on lowering of yield stress was not studied. Values obtained for the circular cylindrical tank were compared with ASME VIII divisions 1 and 2 standard values and found not to be significantly different ($P > 0.05$).

REFERENCES

1. KSP 10-K, "Transportation of Petroleum Products," 2012. www.wikinvest.com/stock/KSea_ Transportation_Partners_ LP_%28KSP%29/Transportation
2. J. O. Ekpen, "Oil Refinery in Nigeria. (Myths and Truths)," NAEE Conference, Abuja, 28-29 April 2011, 2012. http:// naee.org.ng/files/
3. K. Babasoki, "Abuja Tanker Explosion," 2012. Fromogala. wrodpress.com/2011/11/16/latest-news-fromthe-abuja-tanker explosion
4. A. Ola and O. David, "10 Killed in Multiple Tanker Explosions, Building Collapse," 2012. wwww.vanguardngr. com/2012/06/10-killed-in multiple-tanker-explosion-building-collapse
5. Daily Mail Reporter, "86,000 Gallon Oil Tanker Bursts into Flames in Highway Crash Caused by Drunken Driver," 2012. www.wdailymail.co.uk/news/article-2126891/fuel-tanker-bursts-into-flames-us-freeway-crash-caused-by-drunken-driver
6. M. Emeka, "Ogun State Government Silent on Fuel Tanker Explosion," Vanguard, 2012, 6 Pages.
7. O. Uchechukwu and A. Boila, "Vehicles Burnt at Ibadan Tanker Explosion," 2012. www.tribune.com/ng/.../37648-12vehiclesburnt-inibadan-tanker
8. C. Miley, "Melborne Tanker Explosion Burns 9," 2012. news ninemsn.com.all/ national/8387836/petrol-tanker-explodes-in-melborne-suburb

9. B. Godfey and O. Ifeyinwa, "Oil Tanker Explosion Destroys Containers, Cars at Apapa Port," 2012. www.vanguardngr.com/2012/04/oil-tanker explosion-destroys-containers-cars-at-apapa-port
10. L. Z. Josef, R. Franz and S. Sebastian, "Pressure Vessel Design—The Direct Route," 1st Edition' Elsevier ltd., Oxford, 2006.
11. G. B. Richard, "Advance Strength and Applied Stress Analysis," 2nd Edition, McGraw Hill Publishing Companies Inc., New York, 1999.
12. R. B. George, 1995. "Finite Element Analysis," Shaum's Outline Series, McGraw-Hill, New York, 1995, 73 Pages.
13. Đ. Mirko, P. Dragan and B. Milan, "Identification of the Stress-Strain State of a Cylindrical Tank with Walls of Variable Thickness," FME Transactions, Vol. 39, No. 1, 2011, pp. 25-32.
14. V. P. Anatoly, Y. F. Sergiy, S. K. Viktor and Z. K. Edward, "Analysis of the Stressed State of Cylindrical Tanks with Defects of Geometrical Shape," SCAD Group, Kiev, 2003.
15. M. Jaroslav, 1996. "Finite Elements in the Analysis of Pressure Vessels and Piping—A Bibliography (1976- 1996)," International Journal of Pressure Vessels and Pipings, Vol. 69, No. 3, 1996, pp. 273-339.
16. M. Patrick and A. O. Thomos, "Graphics and GUIs with Matlab," Chapman & Hall CRC Press Company, London, 2003.
17. T. Timoshenko and J. Goodier, "Theory of Elasticity," McGraw Hill Books Company Inc., New York, 1951.
18. C. Ross, "Advance Stress Analysis," Ellis Horwood Limited, New York, 1987.
19. J. N. Ready, "An Introduction to Finite Element Method," 3rd Edition, McGraw Hill, New York, 2006.

Chapter 7

Greener Pathways for Energy-Intensive Commodity Chemicals: Opportunities and Challenges

Yuan Yao[1], Diane Graziano[2], Matthew Riddle[3], Joe Cresko[4], and Eric Masanet[1, 5]

[1]Department of Chemical & Biological Engineering, Northwestern University, 2145 Sheridan Road, Evanston, IL 60201, USA

[2]Global Security Sciences Division, Argonne National Laboratory, 9700 South Cass Avenue, Argonne, IL 60439, USA

[3]Energy Systems Division, Argonne National Laboratory, 9700 South Cass Avenue, Argonne, IL 60439, USA

[4]Advanced Manufacturing Office, United States Department of Energy, 1000 Independence Avenue, SW, Washington, DC 20585, USA

[5]Department of Mechanical Engineering, Northwestern University, 2145 Sheridan Road, Evanston, IL 60201, USA

ABSTRACT

The chemical industry is poised for significant growth and investment, which presents an opportunity for adoption of greener chemical technologies. This article reviews available and emerging technologies for reducing the fossil fuel demand associated with the ammonia, ethylene, methanol, propylene, and benzene, toluene, and xylenes (BTX) industries. These few energy-intensive commodity chemicals (EICCs) account for around half of the energy use and greenhouse gas (GHG) emissions of the global chemical industry. Available data are harmonized to characterize potential energy use and GHG emissions savings, while technical and economic barriers to adoption are discussed. This information sheds light on the status of future technological options for reducing the impacts of the chemicals industry, and provides quantitative data to industry analysts and policy makers seeking a greater understanding of such options for EICCs.

INTRODUCTION

As economies recover from the 2008 recession, demand for chemicals is increasing. The global chemical industry is projected to grow by 4.1% in 2015 [1] with continued growth over the coming decades [2]. In the United States, shale gas has improved the competitiveness of the chemical industry and has spurred increased capital investment, production, and exports [3]. By 2018, the American Chemistry Council forecasts capital spending to reach $61 billion in the United States and $618 billion globally [1]. As new plants are constructed and existing plants are expanded and upgraded, opportunities arise for adoption of greener technologies, including more energy efficient equipment, greener solvents [4 and 5], improved reactions and catalysts [6 and 7], alternative feedstocks

[8•• and 9••], and process intensification [10, 11 and 12], to name a few. Such technologies align with the Twelve Principles of Green Chemistry [13], which fall broadly into two categories: first, energy and materials efficiency; and second, human and ecological health protection.

Greener technology options that reduce fossil fuel inputs for process energy and/or feedstocks are particularly important moving forward. Chemical production is the world's largest industrial energy consumer, with a 30% share of global industrial energy use (including feedstocks) [14•]. Reducing fossil fuel inputs can decelerate resource depletion, reduce air pollution, and mitigate environmental stresses related to fossil fuel extraction, transport, and use [15] — benefits that align well with the Twelve Principles. Furthermore, the chemical industry is the world's third largest source of industrial greenhouse gas (GHG) emissions, due largely to fossil fuel combustion for process energy [16]. Therefore, reducing fossil fuel inputs is also a crucial climate change risk mitigation strategy, both for the industry and for society [17].

This article reviews available and emerging technologies for reducing fossil fuel use associated with ammonia, ethylene, methanol, propylene, and benzene, toluene, and xylenes (BTX). As indicated by Figure 1, these few energy-intensive commodity chemicals (EICCs) account for around half of the energy use and GHG emissions of the global chemical industry [18••]. Specifically, this article harmonizes recent data on the performance and barriers associated with more energy-efficient production processes, alternative feedstocks, and substitutes for these EICCs and their derivatives. While this summary is not comprehensive of all possible technology improvements, it focuses on pathways for which recent and credible data exist in the literature related to potential energy and GHG emissions reductions and implementation barriers. Section 'Energy-efficient pathways from fossil fuel feedstocks' includes pathways that reduce process energy demand within EICC plants, while Section 'Alternative feedstock pathways' includes pathways that can reduce or eliminate the need for fossil fuel feedstocks for EICCs and their derivatives.

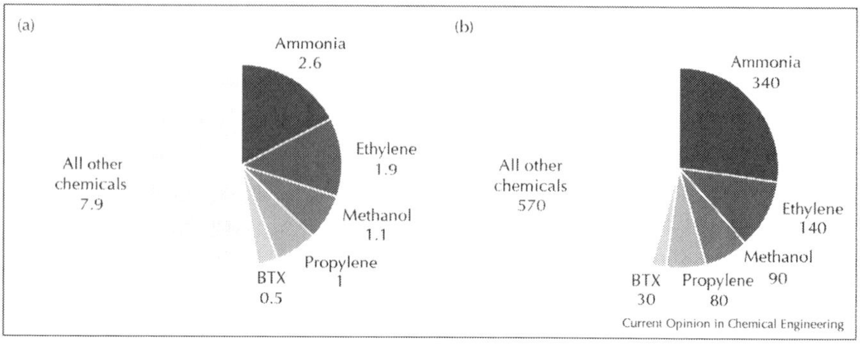

Figure 1: International Energy Agency estimates of the (a) energy consumption (in EJ/year) and (b) GHG emissions (in $MtCO_2$-eq/year) of the global chemical industry in 2010 [18••].

ENERGY-EFFICIENT PATHWAYS FROM FOSSIL FUEL FEEDSTOCKS

Production of EICCs is predominantly based on fossil fuel feedstocks, which will continue for the foreseeable future. Most EICCs are manufactured in highly integrated, multi-chemical plants that have steadily improved energy efficiencies over time, as evidenced by a 39% reduction in the energy intensity of the U.S. chemical sector between 1994 and 2007 [18••]. Common improvements include process heat integration, advanced controls, and best practice technologies for steam, pumps, motors, and other plant systems [14•, 19 and 20]. Still, chemical plant audits regularly identify technology options for even greater efficiency [21 and 22], and the state-of-the-art for many processes continues to evolve [20] Table 1summarizes recent performance data on several energy efficient process technologies for producing ethylene, propylene, and ammonia from current feedstocks, which can reduce direct fossil fuel inputs and GHG emissions compared to prevailing process technologies.

Table 1: Energy efficient technologies for energy intensive chemicals using fossil feedstock

Chemicals	**Technology**	**Feedstock**	**Process description**	**Development stage**	**On-site energy reduction compared to reference technology[a](GJ/t product)**
Ethylene	Catalyst-assisted production of olefins (CAMOL™) [23]	NGL	Cracker catalyst coating reduces coke deposits	Pilot plant test	1–2 (6–10%)
	Membrane/distillation hybrid process [27]	NGL	A technology integrating membrane and distillation	Lab scale	1–1.7 (5–9%)
	Microchannel process technology (MPT) distillation [49]	NGL	A novel process which combines heat transfer and gas separation into one unit	Lab scale	0.6 (3% saving for entire process, 16% saving for distillation units)[b]
	Enhanced separation efficiency in olefin/ paraffin distillation through hollow fiber technology [50]	NGL	Replacement of conventional packing materials with hollow fibers	Lab scale	0.4 (2% saving for entire process, 11% saving for distillation units)[b]
	Ethane oxidative dehydrogenation [30 and 31]	Ethane	Ethane to ethylene through catalytic reaction	Lab scale	6.7–10.3 (35–54%)

Propylene	Propane dehydrogenation (PDH) [30 and 37]	Propane	Directly conversion of propane to propylene and hydrogen	Commercial	5–7 (40–50%)
Ammonia	Topsøe low-energy ammonia technology [51]	Methane	Package includes Haldor Topsøe heat exchanger, better catalyst design, and proprietary converter design	Commercial	6–8 (17–22%)
	KBR KAAPplus [39]	Methane	Package includes ATR parallel with exchanger reformer, ruthenium catalyst, clean and dry make-up gas to the synthesis loop	Commercial	8.6 (24%)
	ATR (autothermal reformer) [52 and 53]	Methane	Partial oxidation and steam reforming are highly integrated in one reactor, which is more efficient than a pure steam reforming process	Commercial	7.2 (20%)
	Reforming exchanger [54]	Methane	Improved heat recovery in primary reformer section	Commercial	≤2.5 (≤7%)
	Process integration of ammonia reactor and ammonia PSA recovery [40 and 55]	Methane	Process integration of ammonia reactor and ammonia pressure swing adsorption	Lab-scale	14–18 (40–50%)

[a]The reference technology for ethylene is ethane cracking process (19 GJ/t ethylene [30 and 56]); for propylene is naphtha cracking process (15 GJ/t propylene [14•]), and for ammonia is ammonia synthesis process (36 GJ/t ammonia [14• and 57]) [27].

Ethylene and Propylene

These two light olefins are currently produced by steam cracking of natural gas liquids (NGL) and naphtha or as byproducts of the petroleum refining process [14•]. The technology data in Table 1 indicate a wide range in potential onsite energy intensity reductions, as well as suitability for increasingly competitive natural gas feedstocks [3].

The five process intensification technologies for ethylene in Table 1 are listed in order of decreasing technological maturity, and all but one are still at the lab scale. Catalyst-assisted production of olefins (CAMOL™) [23] is currently being demonstrated at the pilot scale, with estimated energy intensity reductions attributable to reduced decoking energy use in the furnace coils of the ethane cracker. No data are available on the expected investment costs of CAMOL-coated coils. However, given that ethylene production costs are dominated by the cost of input fuels [3], the substantial reported energy savings are likely to justify the incremental costs for coated coils. Additional pilot tests and verified performance and cost data should lead to a commercialized technology in the next several years.

The hybrid membrane/distillation process has been researched for many years, but has yet to see commercial adoption due to barriers related to application design optimization [24] and reliable prediction of membrane module performance [25]. Available economic analyses indicate significant investment savings of 54–61% compared to a conventional C2-splitter and 11–36% compared to a C3-splitter [26]. Similarly, Compact Membrane Systems reported a return on investment of up to 67% with a payback time of 1.3 years for a hybrid system [27]. However, more research efforts are needed to address the aforementioned process design barriers, and pilot applications are necessary to verify cost and energy savings to bring this technology closer to commercialization.

The advantage of microchannel process technology (MPT) lies in its improved heat and mass transfer performance compared to conventional reactors and distillation units. Currently, its high capital cost, which is mainly attributable to its manufacturing complexity, serves as a major barrier to commercialization and widespread adoption by olefin plants. Given its modularity and flexibility, however, it may find a responsive early market in small scale distillation facilities if capital costs can be lowered through manufacturing process cost reductions.

Hollow fibers as distillation packing materials deliver smaller estimated energy intensity reductions compared to the preceding options, so low investment costs will be crucial to the success of this lab-scale technology. Recent data suggest that hollow fibers can provide stable and efficient olefin/paraffin distillation with high capacities [28]. However, several engineering barriers must be overcome before commercialization, including identification of robust, durable, and economical polymeric fibers and challenges related to efficient scaled-up process design [28]. A future scale-up test will be conducted by Chevron [29], which may help to accelerate the commercialization process.

Alternative chemical pathways being investigated for the production of ethylene include: first, ethane oxidative dehydrogenation (EOD) [30 and 31]; second, oxidative coupling of methane (OCM) [32]; and third, methanol to olefins (MTO) [32 and 33]. At the lab scale, EOD catalysts have improved ethylene yields to 45–75% [34], which is comparable to conventional technologies. Patent data for EOD technology suggest a lower investment cost for separation/recovery, due to reduced product mix complexity compared to conventional steam cracking [35]. While other cost data are not reported, its complex reactor design [36] and intensive catalyst usage [31] may lead to high system costs. Therefore, the future commercialization of this technology likely hinges on a more efficient and economical reactor and more durable and stable catalysts.

OCM and MTO are absent from Table 1 due to lack of conclusive data on process energy intensities compared to conventional cracking. However, one OCM pilot facility is planned in La Porte, Texas for 2015, which will contribute to its development and feasibility [81]. MTO is commercialized and viable in regions where methane is less costly than steam cracking feedstock, but analysis data suggest that MTO has a higher onsite energy intensity than conventional pathways [18•• and 32]. However, if coupled with renewable methanol or renewable energy sources, MTO holds great promise for greener ethylene production.

Propane dehydrogenation (PDH) use is growing rapidly due to the recent shale gas boom, with new PDH plants announced that will range in capacity from 600 to 750 thousand tons (t) of propylene per year [30 and 37]. Furthermore, the production cost of PDH is \$885/ton, which is highly favorable compared to the average U.S. propylene price of \$1428/ton in 2012 [38]. Thus, PDH delivers substantial reductions in both energy intensity and production costs, which makes it an exemplary greener chemistry technology option, especially as propylene demand and propane supply (due to shale gas) increase moving forward. Additionally, H_2 produced by PDH offer additional revenue if markets are nearby.

Ammonia

Ammonia is currently produced by the Haber-Bosch [39] process using methane feedstock. The energy efficient technology options listed for ammonia in Table 1 include four commercial technologies, whose adoption is driven by production characteristics and investment criteria that can vary across plants, and one lab-scale technology. All technologies were included here to illustrate the potential energy savings associated with state-of-the-art equipment upgrades at ammonia plants compared to conventional ammonia synthesis processes. Furthermore, the energy intensity reductions in Table 1 represent the most recent data from the literature harmonized to a consistent comparison basis, which underscores the range of efficiency improvements achievable through different process equipment selections. In particular, the use of packaged designs can lead to energy intensity improvements compared to conventional systems, and such packages can be used to replace existing process equipment or in new lines for expanded production.

The lab-scale ammonia production technology demonstrates that significant energy efficiency improvements might still be had for processes as mature as Haber-Bosch, and for which a number of commercial efficiency options already exist [40]. This process intensification technology leads to estimated energy intensity reductions substantially greater than the commercialized ammonia process technologies. Preliminary cost data suggest slightly lower capital costs than other options as well [40]. However, pilot test data at larger scales are needed to verify the energy savings and cost competitiveness of this technology before its commercialization potential can be assessed.

Methanol and BTX

For methanol and BTX, a particularly promising pathway is from direct conversion of methane. While no energy savings data have been reported for direct conversion pathways, energy savings are expected at production scales due the greatly simplified process inherent in direct conversion routes. Therefore, much research into direct conversion of methane is underway, which may lead to viable process technologies in the near future with continued effort.

Several direct conversion pathways have been reported for methanol [41], including gas-phase catalytic partial oxidation [42]; liquid-phase catalytic oxidation [43 and 44]; and water-based biological oxidation. The latter approach attempts direct conversion of methane to methanol via methane monooxygenases enzymes at ambient temperature. Research on the direct conversion to methanol derivatives, such as formaldehyde [45 and 46] and acetic acid [47]), has also been reported. For BTX, direct non-oxidative conversion of methane to aromatics [41 and 48] has been demonstrated at the lab scale.

Alternative Feedstock Pathways

A large share of fossil fuel use for chemicals (e.g., around 50% in the United States [58]) is attributable to feedstocks. Research on alternative feedstocks is vibrant, primarily due to rapidly growing demand for 'green' chemical products. Quantifying the environmental implications of alternative feedstocks requires a life-cycle assessment (LCA) approach that considers mass and energy inputs and related environmental impacts across the entire product life cycle [59].

This section reviews emerging technologies as well as recently available (but sparse) data from the LCA literature on the use of biomass and byproduct CO_2 as chemical feedstocks for EICCs, their derivatives, and substitute products. While alternative feedstock pathways for EICC derivatives and substitutes will not directly affect the energy/carbon intensity of conventional EICC production, they could lead to reductions in the energy use and GHG emissions of the global chemical sector due to decreased demand for conventional EICCs. Table 2 summarizes the available data on life-cycle savings and current production costs, expressed relative to conventional fossil fuel feedstock pathways for ease of comparison.

Table 2: Published LCA results for biomass and CO_2 pathways associated with the EICCs and their derivatives

Chemical [references]	Conventional process	Feedstock[a]	Conversion pathway	Life cycle energy reduction GJ/t product (%)	Life cycle GHG reduction t CO_2-eq/t product (%)	Development stage	Cost ratio (bio-/fossil-based)
Bio-adipic acid [69 and 75] (benzene derivative)	Petroleum-based, benzene building block	Glucose from sugar beet	Bio-chemical & Thermal-chemical	30.4–58.9 GJ/t (29–57%)	9.8–17.4 tCO_2-eq/t (NA)	Pilot plant, commercialization in 1–2 years	0.8–0.7
Bio-polylactic acid PLA [76 and 77] (substitute)	50% PE, 50% polypropylene (PP) from ethylene, propylene (PLA replace PP and PE as plastics)	Corn starch	Bio-chemical	22–50 GJ/t (29–66%)	0.8–3.0 tCO_2-eq/t (17–63%)	Commercialized	0.9–1.3
Bio-polyethylene terephthalate PET [78] (ethylene, xylene derivative)	Petroleum based PET produced from ethylene glycol and purified terephthalic acid/dimethyl terephthalate	Maize and sugarcane	Bio-chemical & thermo-chemical	13–22 GJ/t (24–32%)	1–2 tCO_2-eq/t PET (41–43%)	Commercialized	Comparable price with petroleum-based PET

Bio-ethylene [67 and 71]	Steam cracking of NGL/naphtha/gas oil	Sugarcane, lignocelluloses, maize starch	Bio-chemical	40–100%	40%	Commercialized	1.1–2.3
Bio-polyethylene [8••] (ethylene derivative)	Polyethylene (PE) from ethylene	Corn starch or sugar cane	Bio-chemical	29.3–67.6 GJ/t (40–88%)	2.1–4.2 tCO_2-eq/t (120–200%)	Commercialized	Depends on bio-ethylene production
Bio-polyhydroxy-alk-anoate (PHA) [8••, 67 and 77] (substitute)	Polyethylene (PE) from ethylene (PHA replace PE as biodegradable plastics)	Agriculture residues, corn starch, sugar cane, lignocelluloses	Bio-chemical	−35 to 58.9 GJ/t (−47% to 77%)	−2.6 to 2.8 tCO_2-eq/t (−160% to 175%)[b]	Commercialized	4.9–5.4
Methanol [79]	Methane	CO_2 captured from power plant; CO generated by thermochemical splitting of CO_2 using solar thermal energy; H_2 generated by water gas shift of H_2O and CO	Methanol synthesis from CO/H_2	10 kWh/kg (98%)	2.4 tCO_2-eq/t (350%)[c]	Commercialized	2.7

Methanol [80]	Methane	CO_2 captured from power plant; H_2 generated by electrolysis supplied by wind farm	Direct methanol synthesis from CO_2/H_2	NA	1.1 tCO_2-eq/t (59%)		

[a]As reported in the source study.

[b]Negative values indicate emerging pathway is more energy and/or GHG intensive than the conventional pathway; percent savings greater than 100% are attributable to avoided emissions.

[c]Large GHG emissions reductions are attributable to carbon capture and renewable energy.

Biomass

As illustrated in Figure 2, biomass can be converted into chemicals through either biochemical or thermo-chemical pathways [60]. The former approach uses biocatalysts, enzymes, bacteria, or other microorganisms to convert biomass to fuels and chemicals [61], while the latter uses heat to break down biomass into hydrocarbons that are further processed into EICCs and their derivatives.

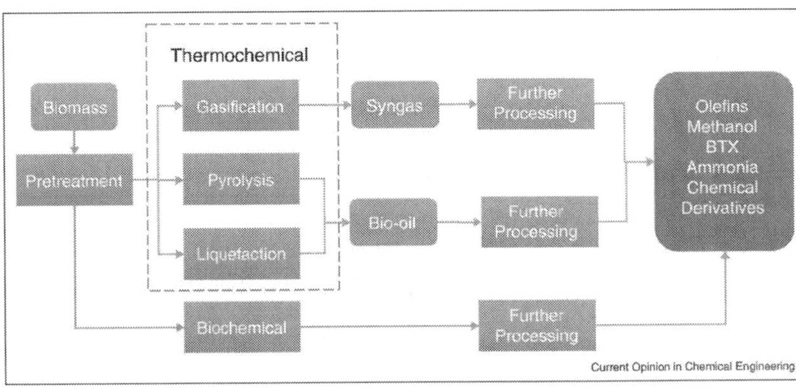

Figure 2: Biomass to chemicals pathways [60].

Most commercialized applications for biomass feedstocks in the literature are limited to high value chemicals and polymers, and at production scales much smaller than conventional EICC plants [62•]. One exception is bioethylene, for which plants in Brazil and India represent about 0.3% of global ethylene capacity using catalytic dehydration of bioethanol from sugarcane [63]. Future development of biorefineries that integrate both biofuel and bio-based chemical production could be transformational in advancing these pathways due to co-product synergies and improved economies of scale [63]. The reader is referred to [62•] for an excellent review of the commercial status and outlook for the production of EICC or their derivatives from biomass feedstocks.

The seven biomass feedstock pathways summarized in Table 2 include commercialized and pilot scale technology pathways for ethylene and derivatives or substitutes for ethylene, propylene, and BTX. Included in these pathways are diverse sources such as agricultural residues, plant oils, and commodity crops, which underscore the broad diversity of emerging biomass feedstock applications. Also included are ranges of costs estimates from the available literature, expressed relative to costs of conventional fossil-fuel-based chemical production.

The identified data in Table 2 suggest that substantial life-cycle energy and GHG emissions savings might be achievable through the adoption of biomass feedstock pathways for EICCs and their substitutes and derivatives. In most cases, savings are primarily attributable to less energy-intensive process pathways and carbon cycling benefits of biomass use. However, it must be noted that, as with all LCAs, significant uncertainties are associated with the source data. For example, previous LCA studies indicate large uncertainties [64, 65 and 66] on results for biofuels due to the complexities of their life-cycle systems. In particular, the identified range of results for bio PHA suggest that life-cycle savings are not guaranteed for all system variations [8•• and 67]. Still, compared to Table 1, the data in Table 2 suggest that biomass feedstocks pathways may hold promise for reducing the energy footprint (GJ/t) of EICC production beyond what may be achieved through improved process efficiencies in prevailing fossil fuel feedstock pathways alone.

Among all bio-based chemicals listed in Table 2, bio-based PLA is the only chemical whose cost benefits has been validated through its application in packaging, coatings, printing and medical implants [68]. Bio-based adipic acid is very close to commercialization and is reportedly cost competitive [69]. While no verifiable cost data could be found for bio-based PET, it is expected to be reasonably cost-competitive compared to petroleum-based PET due to its adoption in commercial soft drink bottling applications [68 and 70].

In addition to technology immaturity and challenges with scaling up to large production quantities, feedstock and processing costs are another major risk factor associated with bio-based chemicals. For example, the cost of bio-ethylene production depends heavily on its region, with higher costs presently reported in the United States ($2000/t from corn) and Europe ($2600/t from sugar beets) and substantially lower costs reported in Brazil and India ($1200/t) [71]. To overcome these significant cost barriers, substantial improvements will be needed throughout the value chain, including increased yields and greater logistical efficiencies for feedstocks, improved efficiencies and cost reductions in conversion processes, and more value placed on bio-based chemicals through increased consumer demand, greater public subsidies to the industry, or both.

Byproduct Carbon Dioxide

While CO_2 is an established feedstock in chemicals production (e.g., for urea and salicyclic acid), there is increasing focus on utilization of waste CO_2 as a strategy for GHG emissions mitigation. Unlike CO_2 that is purposely produced for chemicals, waste CO_2 refers to CO_2 that is currently discharged to the atmosphere as a combustion byproduct (e.g., from boilers, furnaces, or electric power plants). The reader is referred to [72] for a comprehensive review of chemical pathways for recovered waste CO_2, which range from hydrogenation for the production of methanol (existing) to electrocatalytic or photocatalytic reduction of CO_2 for the production of olefins (exploratory).

Table 2 includes LCA data for producing methanol from recovered waste CO_2 and hydrogen produced using renewable power, which are situations that can lead to substantial life-cycle GHG emissions savings [72 and 73•]. The high costs of such processes, which are mostly attributable to hydrogen production, currently present a major barrier to wider adoption. If a cheaper route of hydrogen supply is available (e.g., hydrogen produced as a byproduct from PDH plants, as discussed above), the cost of using waste CO_2 for methanol might be reduced significantly.

Still, the use of renewable resources is crucial as the economic and environmental benefits of these methanol pathways would be offset by energy intensive H_2 production and CO_2 purification processes. The identified data shed light on one example of a shift to a 'methanol economy,' which has been espoused by thought leaders to promote carbon recycling by capitalizing on the utility of methanol as both a fuel and a chemical building block [73•].

SUMMARY

The data summarized here shed light on the potential of available and emerging technologies for reducing fossil fuel dependencies associated with EICCs and their substitutes and derivatives. For available and mature technologies, barriers are often related to limited budgets for capital investments — even when life-cycle costs are favorable — but increased plans for production expansions over the next few years might help overcome these barriers for many plants. Additional barriers include risk aversion, sunk capital costs in existing equipment, and knowledge barriers related to efficient technology options and their expected performance [74]. For the emerging technologies and pathways, much research and investment effort is still needed to improve both technical performance and cost competitiveness, and accelerated pilots and demonstrations will be crucial for commercialization and market uptake in the near term.

While this review characterized technologies based on available literature data, there are certainly more emerging greener chemistry technologies for EICCs at the lab and pilot scales for which performance data do not exist in the public domain. Furthermore, uncertainties associated with published performance data and analyses limited the scope of this article to technologies for which only substantial energy use and emissions reductions are expected. There is a pressing need for more credible analysis data on such technologies, so that chemicals industry engineers, energy analysts, and policy makers can identify and assess future technology options as society seeks to minimize its environmental footprint. That said, the information presented here sheds some credible light on the status of future technological options for reducing the impacts of the chemicals industry, and provides quantitative data to industry analysts and policy makers seeking a greater understanding of such options for EICCs for deeper analyses of their own.

ACKNOWLEDGEMENTS

This work was supported by DOE under contract number DE-AC02-06CH11357. The data and views expressed in this paper are those of the authors and are not endorsed by the U.S. Department of Energy or the United States government.

REFERENCES

1. American Chemistry Council: Year-end 2013 Chemical Industry Situation and Outlook. Washington, DC, USA: American Chemistry Council; 2013,
2. Valencia RC: The Future of the Chemical Industry by 2050. WileyVCH; 2013.
3. Coopers PW: Shale gas: reshaping the US chemicals industry. 2012.
4. Simon M-O, Li C-J: Green chemistry oriented organic synthesis in water. Chem Soc Rev 2012, 41:1415-1427.
5. Gu Y, Jerome F: Glycerol as a sustainable solvent for green chemistry. Green Chem 2010, 12:1127-1138.
6. Sheldon RA: Fundamentals of green chemistry: efficiency in reaction design. Chem Soc Rev 2012, 41:1437-1451.
7. Li C-J, Trost BM: Green chemistry for chemical synthesis. Proc Natl Acad Sci 2008, 105:13197-13202.
8. Chen G-Q, Patel MK: Plastics derived from biological sources: present and future: a technical and environmental review. Chem Rev 2011, 112:2082-2099. This paper reviewed recent R&D developments and LCA studies (life cycle assessments) of bio-plastics. By comparing bio-based plastics with petroleum-derived plastics from both cost and environmental

perspectives, the authors presented several attractive bioplastic products and pointed out where there is room to reduce the energy consumption and GHG emissions of bioplastics if they are put into large scale production.

9. Bozell JJ, Petersen GR: Technology development for the production of biobased products from biorefinery carbohydrates — the US Department of Energy's "Top 10" revisited. Green Chem 2010, 12:539-554. This paper presents a updated version of a report published by the U.S. Department of Energy in 2004 on the "top 10" chemicals that have the largest potentials to be intergrated into biorefinery's portfolio in a economically viable and technically feasible way. The authors evaluate the potentials of different chemicals from nine perspectives and discuss the technology developments relating to each chemical.
10. Vaccaro L, Lanari D, Marrocchi A, Strappaveccia G: Flow approaches towards sustainability. Green Chem 2014, 16:3680-3704.
11. Newman SG, Jensen KF: The role of flow in green chemistry and engineering. Green Chem 2013, 15:1456-1472.
12. Moseley JD, Kappe CO: A critical assessment of the greenness and energy efficiency of microwave-assisted organic synthesis. Green Chem 2011, 13:794-806.
13. Anastas PTA, Warner JCA: Green Chemistry: Theory and Practice. Oxford University Press; 2000.
14. IEA: Chemical and Petrochemical Sector, Potential of Best Practice Technology and Other Measures for Improving Energy Efficiency. France: IEA; 2009, . This report analyzes worldwide potentials for energy savings and GHG emissions reductions in the chemical industry through the adoption of best practice technologies. Results indicate a large reduction potential in energy use, especially within the petrochemical industry, but also highlight how more efforts and progress are needed for data availability and quality. This report is not only a good resource for energy intensity data of chemical

production, but it is also an insightful analysis showing technology differences between the chemical industries of different countries.

15. Masanet E, Chang Y, Gopal AR, Larsen P, Morrow WR, Sathre R, Shehabi A, Zhai P: Life-cycle assessment of electric power systems. Annu Rev Environ Resour 2013, 38:107-136.
16. Sabine Brueske RS, Zach C, Andres H: U.S. Manufacturing Energy Use and Greenhouse Gas Emissions Analysis. Oak Ridge National Laboratory; 2012.
17. IPCC: Climate Change 2014: Impacts, Adaptation, and Vulnerability. Switzerland: IPCC; 2014,
18. IEA, ICCA, DECHEMA: Technology Roadmap, Energy and GHG Reductions in the Chemical Industry via Catalytic Processes. France: IEA, ICCA, DECHEMA; 2013, This report evaluates the potential energy savings and GHG emissions reduction potentials that can be realized by continuous improvements and breakthrough technology options in the both short to medium term (2025) and long term (2050). In order to make these projections, this analysis identified 18 chemical products that accounts for 80% of energy demand and 75% of GHG emissions in the global chemical industry; and looked at the best practice technologies, incremental improvements and emerging technologies for these 18 chemicals.
19. Neelis ML, Worrell E, Masanet E: Energy Efficiency Improvement and Cost Saving Opportunities for the Petrochemical Industry. Berkeley, CA, USA: Lawrence Berkeley National Laboratory; 2008, .
20. ENERGETICS: Chemicals Industry Updated Energy Bandwidth Study. U.S. Department of Energy; 2013.
21. Masanet E, Walker ME: Energy-water efficiency and U.S. industrial steam. AIChE J 2013, 59:2268-2274.
22. Wright A, Martin M, Gemmer B, Scheihing P, Quinn J: Results from the U.S. DOE 2006 Save Energy Now Assessment Initiative. Washington, DC: U.S. Department of Energy; 2007,

23. BASF: CAMOLTM Catalytic Coatings for Steam Cracker Furnace Tubes. Canada: BASF; 2012,
24. Kookos IK: Optimal design of membrane/distillation column hybrid processes. Ind Eng Chem Res 2003, 42:1731-1738.
25. Bernardo P, Drioli E, Golemme G: Membrane gas separation: a review/state of the art. Ind Eng Chem Res 2009, 48:4638-4663.
26. Benali M, Aydin B: Ethane/ethylene and propane/propylene separation in hybrid membrane distillation systems: optimization and economic analysis. Sep Purif Technol 2010, 73:377-390.
27. Program SBIRS: Novel Membranes for Olefin/Paraffin Separation. SBIR/STTR; 2011.
28. Yang D, Le L, Martinez R, Morrison M: Hollow fibers structured packings in olefin/paraffin distillation: apparatus scale-up and long-term stability. Ind Eng Chem Res 2013, 52:9165-9179.
29. DOE: Enhanced Separation Efficiency in Olefin/Paraffin Distillation. US DOE; 2011.
30. Ren T, Patel M, and Blok K: Olefins from conventional and heavy feedstocks: energy use in steam cracking and alternative processes. Energy 2006, 31:425-451.
31. Cavani F, Ballarini N, Cericola A: Oxidative dehydrogenation of ethane and propane: how far from commercial implementation? Catal Today 2007, 127:113-131.
32. Ren T, Patel MK, and Blok K: Steam cracking and methane to olefins: energy use. CO2 emissions and production costs. Energy 2008, 33:817-833.
33. Chen JQ, Bozzano A, Glover B, Fuglerud T, Kvisle S: Recent advancements in ethylene and propylene production using the UOP/hydro MTO process. Catal Today 2005, 106:103-107.
34. Agouram S, Dejoz A, Ivars F, Va´ zquez I, Lo´ pez Nieto JM, Solsona B: Oxidative dehydrogenation of ethane: a study over the structure and robustness of Ni–W–O catalysts. Fuel Process Technol 2014, 119:105-113.

35. Arnold SC, Gaffney AM, Song R, Yeh CY: Process for producing ethylene via oxidative dehydrogenation (odh) of ethane. US Patent 2010.

36. Ga¨rtner CA, van Veen AC, Lercher JA: Oxidative dehydrogenation of ethane: common principles and mechanistic aspects. ChemCatChem 2013, 5:3196-3217.

37. Intratec Solutions: Technology Economics: Propylene via Propane Dehydrogenation, Part 2. CreateSpace Independent Publishing Platform; 2013.

38. Horncastle A, Sastry A, Corrigan J, Gotpagar J: Future of Chemicals Rebalancing Global Feedstock Distruptions with "Onpurpose" Technologies. Booz & Company; 2012.

39. Appl M: Ammonia. Ullmann's Encyclopedia of Industrial Chemistry, Electronic Release. Wiley-VCH; 2010.

40. SmartKoncept Technology: Final Scientific Technical Report: Ammonia Process by Pressure Swing Adsorption. Houston, TX, USA: SmartKoncept Technology; 2010, .

41. Alvarez-Galvan MC, Mota N, Ojeda M, Rojas S, Navarro RM, Fierro JLG: Direct methane conversion routes to chemicals and fuels. Catal Today 2011, 171:15-23.

42. Shan J-J, Huang W, Nguyen L, Yu Y, Zhang S, Li Y, Frenkel AI, Tao F: Conversion of methane to methanol with a bent mono (m-oxo) dinickel anchored on internal surface of micropores. Langmuir 2014, 30:8558-8569.

43. Hammond C, Forde MM, Ab Rahim MH, Thetford A, He Q, Jenkins RL, Dimitratos N, Lopez-Sanchez JA, Dummer NF, Murphy DM et al.: Direct catalytic conversion of methane to methanol in an aqueous medium by using copper-promoted Fe-ZSM-5. Angew Chem Int Ed 2012, 51:5129-5133.

44. Li T, Wang SJ, Yu CS, Ma YC, Li KL, Lin LW: Direct conversion of methane to methanol over nano-[Au/SiO2] in [Bmim]Cl ionic liquid. Appl Catal A Gen 2011, 398:150-154.

45. Zhang J, Burkle´-Vitzthum V, Marquaire PM, Wild G, Commenge JM: Direct conversion of methane in formaldehyde

at very short residence time. Chem Eng Sci 2011, 66:6331-6340.

46. Wang Z-C, Dietl N, Kretschmer R, Ma J-B, Weiske T, Schlangen M, Schwarz H: Direct conversion of methane into formaldehyde mediated by [Al2O3] + at room temperature. Angew Chem Int Ed 2012, 51:3703-3707.
47. Phan A, Czaja AU, Ga´ndara F, Knobler CB, Yaghi OM: Metal–organic frameworks of vanadium as catalysts for conversion of methane to acetic acid. Inorg Chem 2011, 50:7388-7390.
48. Guo X, Fang G, Li G, Ma H, Fan H, Yu L, Ma C, Wu X, Deng D, Wei M et al.: Direct, nonoxidative conversion of methane to ethylene, aromatics, and hydrogen. Science 2014, 344:616-619.
49. Fanelli M, Arora R, Tonkovich A, Marco J, Rode E: Advanced Distillation Final Report. Plain City, OH, USA: DOE; 2010, .
50. Yang D, Martinez R, Fayyaz-Najafi B, and Wright R: Light hydrocarbon distillation using hollow fibers as structured packings. J Membr Sci 2010, 362:86-96.
51. Haldor Topsøe: Topsøe's Ammonia Technology Today. Haldor Topsoe; 2013.
52. Rice SF, Mann DP: AutothermalReforming of Natural Gas to Synthesis Gas. Livermore, CA: USA Sandia National Laboratories; 2007, .
53. Martin N, Worrell E, Ruth M, Price L, Elliot RN, Shipley AM, Thorne J: Emerging Energy-efficient Industrial Technologies. Lawrence Berkeley National Laboratory; 2000.
54. Singh S, Price J: KRES-ESTM revamp makes more ammonia from less natural gas by energy substitution. Nitrogen + Syngas International Conference & Exhibition; Du¨sseldorf, Germany: 2011.
55. DOE: Grand Challenge Portfolio: Driving Innovations in Industrial Energy Efficiency. Washington, DC, USA: DOE; 2011,.

56. ENERGETICS, DOE: Energy and Environmental Profile of the U.S. Chemical Industry. Columbia, MD: ENERGETICS, DOE; 2000,.
57. Worrell E, Phylipsen D, Einstein D, Martin N: Energy Use and Energy Intensity of the U.S. Chemical Industry. Berkeley, California: Ernest Orlando Lawrence Berkeley National Laboratory; 2000,.
58. EIA: Manufacturing Energy Consumption Survey. Washington, DC: U.S EIA; 2013, .
59. Rebitzer G, Ekvall T, Frischknecht R, Hunkeler D, Norris G, Rydberg T, Schmidt WP, Suh S, Weidema BP, and Pennington DW: Life cycle assessment: part 1: framework, goal and scope definition, inventory analysis, and applications. Environ Int 2004, 30:701-720.
60. Menon V, Rao M: Trends in bioconversion of lignocellulose: biofuels, platform chemicals & biorefinery concept. Progr Energy Combust Sci 2012, 38:522-550.
61. Balat M: Production of bioethanol from lignocellulosic materials via the biochemical pathway: a review. Energy Convers Manage 2011, 52:858-875.
62. Bioenergy I: Bio-based Chemicals — Value Added Products from Biorefineries. The Netherlands: Bioenergy I; 2012, this report provides a comprehensive review of biorefinery developments (both in research and commercialization) in recent years. The authors introduce different chemical platforms that have the potential to be used in biorefineries for large scale chemical production, present possible environmental and economic benefits bought by co-production of fuels and chemicals in a biorefinery, list bio-chemical products commercialized or near market, and point out future challenges and research needs.
63. Clark JH, Luque R, Matharu AS: Green chemistry, biofuels, and biorefinery. Annu Rev Chem Biomol Eng 2012, 3:183-207.

64. Cherubini F, Jungmeier G: LCA of a biorefinery concept producing bioethanol, bioenergy, and chemicals from switchgrass. Int J Life Cycle Assess 2010, 15:53-66.
65. Kajaste R: Chemicals from biomass — managing greenhouse gas emissions in biorefinery production chains — a review. J Clean Prod 2014, 75:1-10.
66. Cherubini F, Ulgiati S: Crop residues as raw materials for biorefinery systems — a LCA case studies. Appl Energy 2010, 87:47-57.
67. Patel MK, Crank M, Dornburg V, Hermann B, Roes L: Medium and Long-term Opportunities and Risks of the Biotechnological Production of Bulk Chemicals from Renewable Resources. Netherlands Utrecht University; 2006.
68. Chen G-Q: A microbial polyhydroxyalkanoates (PHA) based bio- and materials industry. Chem Soc Rev 2009, 38:2434-2446.
69. Boussie T: Scale-up and Commercialization of Bio-based Adipic Acid and HMD. GTC; 2013.
70. Harrington R: Coca-Cola Targets 100% Plant-based Bottle by 2017. 2011.
71. IEA-ETSAP, IRENA Technology: Production of Bio-ethylene. IEAETSAP and IRENA Technology; 2013.
72. Peters M, Ko¨ hler B, Kuckshinrichs W, Leitner W, Markewitz P, Mu¨ ller TE: Chemical technologies for exploiting and recycling carbon dioxide into the value chain. ChemSusChem 2011, 4:1216-1240.
73. Olah GA: Towards oil independence through renewable methanol chemistry. Angew Chem Int Ed 2013, 52:104-107. In this paper, the author presents the concept of a "Methanol Economy" that uses renewable methanol as both fuels and chemical feedstock to replace petroleum oil and other fossil fuels. The author reviews recent technology developments on both methanol applications and renewable methanol production and provides insights on the realization of a methanol economy in the future.

74. Russell C: Barriers to industrial energy cost control: the competitor within. Chem Process 2005.
75. van Duuren JBJH, Brehmer B, Mars AE, Eggink G, dos Santos VAPM, Sanders JPM: A limited LCA of bio-adipic acid: manufacturing the nylon-6,6 precursor adipic acid using the benzoic acid degradation pathway from different feedstocks. Biotechnol Bioeng 2011, 108:1298-1306.
76. Crank M, Marscheider-Weidemann F, Schleich J, Hu¨ sing B, Angerer G: Techno-economic Feasibility of Large-scale Production of Bio-based Polymers in Europe. Wolf O. Spain: DG JRC; 2005, .
77. Saygin D, Gielen DJ, Draeck M, Worrell E, Patel MK: Assessment of the technical and economic potentials of biomass use for the production of steam, chemicals and polymers. Renew Sustain Energy Rev 2014, 40:1153-1167.
78. Shen L, Worrell E, Patel MK: Comparing life cycle energy and GHG emissions of bio-based PET, recycled PET, PLA, and man-made cellulosics. Biofuels Bioprod Biorefin 2012, 6:625- 639.
79. Kim J, Henao CA, Johnson TA, Dedrick DE, Miller JE, Stechel EB, Maravelias CT: Methanol production from CO2 using solarthermal energy: process development and techno-economic analysis. Energy Environ Sci 2011, 4:3122-3132.
80. von der Assen N, Jung J, Bardow A: Life-cycle assessment of carbon dioxide capture and utilization: avoiding the pitfalls. Energy Environ Sci 2013, 6:2721-2734.
81. Siluria Technologies: Siluria Technologies Announces Construction and Site Selection for Its OCM Demonstration Unit in La Porte, Texas, http://siluria.com/Media/ Press_ Releases?page=2. 9

Chapter 8

Petroleum Geochemical Proxies for Reservoir Engineering Parameters

B. Bennett[a], A. Lager[b], D.K. Potter[c], J.O. Buckman[c], and S.R. Larter[a, b]

[a]Petroleum Reservoir Group (PRG), Department of Geology and Geophysics, University of Calgary, 2500 University Drive NW, Calgary, Alberta, Canada T2N 1N4

[b]NRG: School of Civil Engineering and Geosciences, Drummond Building, The University, Newcastle upon Tyne, NE1 7RU, UK

[c]Institute of Petroleum Engineering, Heriot-Watt University, Riccarton, Edinburgh, EH14 4AS, UK

ABSTRACT

The prediction of fluid flow behaviour in petroleum reservoirs is influenced by the physical and chemical processes active in interacting crude oil/brine/rock systems. It is usually not possible to assess these complex systems directly so proxies for molecular scale behaviour are needed. By their very nature, polar non-hydrocarbons are sensitive to fluid–rock interactions, and if properly exploited they may be utilised as proxies for describing reservoir engineering properties (e.g. wettability) that are also sensitive to fluid–rock interactions. We have identified a group of aromatic oxygen (alkylphenols and alkylfluorenones) and aromatic nitrogen (alkylcarbazoles) compounds present in petroleum that appear to respond to variations in fluid–rock properties. Here we describe the chemical and physical changes in a series of core samples obtained from North Sea reservoirs. A number of petrophysical parameters displayed strong correlations with polar non-hydrocarbon occurrence. For example, deflections in gamma ray logs in response to clay content in a coarsening upwards sandstone unit also showed similar deflections from a number of geochemical logs.

A core-flood experiment was designed to monitor the chemical and physical changes during oil migration in a siltstone core. Following completion of the core-flood experiment, Environmental Scanning Electron Microscopy (ESEM) analysis of core samples indicated hydrophilic and hydrophobic surface tendencies grading throughout the core. The distributions of polar non-hydrocarbons (e.g. C0–C3–phenols) appear to correspond closely to the observed wettability alteration. The results confirm the potential for developing proxies for fluid–rock interactions through monitoring the surface active compounds present in the polar non-hydrocarbon fraction of petroleum.

INTRODUCTION

Reservoir geochemistry involves the study of compositional variations of petroleum reservoir fluids (waters, oils and gases) at a variety of spatial and temporal scales, and reveals information about petroleum basin development, the details of reservoir filling and leaking, and petroleum mixing and alteration (Larter et al., 1997). Reservoir geochemistry provides a natural link with reservoir engineering but until now has been relatively poorly exploited. Compositional variations occurring within a petroleum reservoir reflect mainly the geological setting of the reservoir and the way in which migrating fluids arrive and fill the reservoir. The fluids may then redistribute compositionally by gravity driven flow and by diffusion (England, 1990), although some inherited compositional variations may persist to the present day. The charge history of the reservoir and poor overall mixing may be the dominant overall control on fluid compositional variations (Stainforth, 2004). The relationship between certain engineering parameters (gas/oil ratio and reservoir fluid density), reservoir rock properties and fluid geochemistry may be exploited to improve characterisation of petroleum reservoirs by combining geochemical and engineering data (England, 1990).

Reservoir wettability is determined by complex interface boundary conditions acting within the pore space of sedimentary rocks, namely crude oil/brine/rock interactions. These conditions have a dominant effect on interface movement and associated oil displacement. There is much debate over the physical meaning of wettability determinations, but it is generally agreed that the distribution of wetting and of oil- and water-wetted phases in most reservoirs is complex, with most reservoirs having mixed wettability with wettability locally being very variable in nature from pore to pore (Treiber et al., 1972), so called heterogeneous wettability.

Petroleum contains numerous surface active compounds that are sensitive to interactions with minerals and organic matter. The distribution of rock sensitive compounds such as benzocarbazoles (Larter et al., 1996) and alkylfluorenones (Bennett and Larter, 1998) often reveal compositional heterogeneity on a sub-meter scale in reservoir core studies, implying this group of compounds may be developed to probe the nature of the fluid–rock environment throughout sampled reservoir intervals. Larter and Aplin (1995) interpreted meter scale variations in nitrogen compound distributions as potential indicators of variations in wettability, variations reflecting the contrast between a mobile fluid phase versus a component strongly sorbed onto solid phases. Stoddart et al. (1995) showed that the distribution of nitrogen compounds in a core from Eldfisk reservoir (North Sea) was very heterogeneous on a meter scale, the core extract petroleum nitrogen fractions probably being out of equilibrium with the mobile oil.

Cuiec (1977) suggested that the organic nitrogen, sulphur and oxygen compounds (referred to as NSO compounds) may affect reservoir wettability by adsorption onto mineral surfaces. Morrow (1990) also suggested that adsorption of polar compounds from crude oil plays a critical role in determining the wetting properties of reservoir-rock surfaces. Anderson (1986) reviewed the reservoir wettability literature and showed that wettability alteration is a complex process with adsorption of polar compounds and/or deposition onto surfaces of organic material originally present in crude oil. It is commonly accepted that the interfacially active components of crude oil are mainly prevalent in the high molecular weight, polar fractions, namely the resins and asphaltenes (Cuiec, 1984, Buckley et al., 1997 and Buckley et al., 1998) despite these compounds being quite hydrophobic and not easily penetrating water films to reach mineral surfaces. A review of the wettability literature shows a lack of studies involving detailed characterisation of the surface active compounds that may be important contributors to wettability alteration. Such studies are usually limited to the correlation of bulk compositional data, such as total nitrogen content or total sulphur content with wetting behaviour (Akhlaq et al., 1996).

While it is generally assumed that non-hydrocarbons are adsorbed onto reservoir surfaces in nature, there are few unequivocal observations to support this. Mitchell et al. (1990) performed surface chemical analyses of a suite of oil-bearing reservoir cores with varying Amott wettabilities. Increased oil-wetting indices were observed in samples, the surfaces of which had higher carbon and nitrogen contents, implicating selective adsorption of petroleum species onto the surfaces of the reservoir.

Although the relationship between reservoir wettability and gross surface chemistry was clearly demonstrated, and great care was taken to remove extraneous oil, it is generally difficult to be certain that stored cores are representative of the subsurface state. This is because resin or asphaltenic material may precipitate onto the reservoir surfaces during core recovery, natural drying and storage, even when special precautions are taken. Cuiec, 1977 and Cuiec, 1984 core restoration and wettability studies provide further evidence for the adsorption of petroleum species onto water-wet reservoir surfaces on a time-scale of months. Equilibria between petroleum species and reservoir rock components are not well understood. The actual distributions of wetting and variously wetted phases cannot be easily examined at the micron scale at which relevant heterogeneity may exist, but some considerations of the processes that occur during the early filling of a reservoir can be made. The fluid in contact with both mineral and solid organic phases in the rock is assumed initially to be water, so that direct contact between oil and potentially oil-wet phases, such as solid organic matter (Anderson, 1986) will develop later once water films have been ruptured. Wettability alteration has often been ascribed to the adsorption of asphaltenes, which is further supported by the observation that de-asphaltened crude oils no longer exhibited adhesion (Buckley et al., 1998). However, the solubility of asphaltenes in water is so low that it seems unlikely that transport of asphaltenes through an initial water film is part of the process. Kaminsky and Radke (1997) performed calculations to show that asphaltene adsorption on reservoir rock in the presence of a finite water film is not significant, and requires that water-film rupture followed by direct deposition of crude oil onto the rock is the likely origin of wettability reversal (water wet to oil-wet changes in reservoir rock). Recent advances in separation methods targeting the NSO compounds in petroleum now provide the geochemist with the means to develop correlations with reservoir engineering parameters (Bennett et al., in press). In this paper, we describe the behaviour of petroleum geochemical parameters compared to the changes in rock petrophysics and reservoir wettability.

EXPERIMENTAL

Sample information

North Sea reservoir sample suites — A number of reservoir core samples were made available from North Sea oilfields. Sample set-1 consisted of forty one core samples obtained from a typical upper shoreface para-sequence with a basal coarsening upwards sandstone unit (5 ft) with the following rock properties: porosity 17.4–20.6%; permeability 5–69 milliDarcies (mD). The coarsening upwards unit is overlain by a clean sandstone unit (19 ft) with porosities of 11.7–21.1% and permeabilities of 428–1240 mD. Sample strategy was optimised towards obtaining mud contamination-free cores, although an oil based drilling mud was used for cutting the original core. Core plugs used for geochemical analysis were cut from the core using water as lubricant. Sample set-2 consisted of a part of an upper to middle shoreface sequence-facies sequence of clean sandstones (porosity, 17.5–24.5%; permeability, 24–1415 mD) overlying a lower permeability micaceous sandstone (porosity, 13.0–19.9%; permeability, 0.36–47 mD). Twenty representative core sample chips were collected. In the cored section we had access to other petrophysical data from the standard core plugs, and the associated wireline log data, including spectral gamma ray logs. Samples set-3 and set-4 were obtained from Palaeocene reservoirs and consisted of petroleum stained sands. X-ray diffraction analysis was performed on the latter sample sets.

A core-flood experiment was undertaken on a Carboniferous aged siltstone core (89.7 cm long and 5.09 cm diameter) obtained from the West Midlands (U.K.). Physical properties of the core are summarised as follows: porosity, 12%; permeability, 0.01–0.02 mD; total organic carbon content, 0.4% and clay content, 37% (estimated by X-ray diffraction). The oil flood commenced following saturation of the siltstone core (initial water saturated state prior to experiment) with a synthetic North Sea formation water. A typical 41° API gravity, non-biodegraded North Sea black oil was flowed at a rate of 2.4 ml per day (0.1 ml h− 1) for 89 days under typical subsurface conditions i.e., 70 °C, 48.3 MPa (7000 psi). Further details of such a core-flood experiment are described in Larter et al. (2000) and Bennett et al. (2004). On completion of the experiment, 1 cm thick (5 cm diameter) sections were cut from the core and quartered into approximately 15 gram portions.

Extraction of the Core Samples

Core samples (ca. 15 g) were first powdered using a TEMA mill and then they were extracted with an azeotropic mixture of dichloromethane (CH_2Cl_2): methanol (93:7 v/v) for 8 h in a Soxtherm apparatus. The extract was then transferred to a column packed with anhydrous Na_2SO_4 to remove any trace of water. Extract weights were determined following concentration and drying under a stream of nitrogen.

Recovery of Hydrocarbons and Polar Non-Hydrocarbons by Solid Phase Extraction

An aliquot of extract (100 mg) was dissolved in n-hexane (100 µl) and transferred to a C18 non-endcapped (NEC) solid phase extraction cartridge (Jones Chromatography, UK). Internal standards, squalane, 1,1-binaphthyl, fluorene-d10 and carbazole-d8, were added to the solid phase extraction cartridge. Following sample application to C18 NEC solid phase extraction, the hydrocarbon and nitrogen compound containing fractions were obtained using the method as described in Bennett et al. (2002).

The separation of the hydrocarbon fraction isolated during C18 NEC solid phase extraction, into saturated and aromatic hydrocarbon fractions, was carried out using the silver nitrate–silica gel solid phase extraction method described in Bennett and Larter (2000). The three fractions were analysed by combined gas chromatography-mass spectrometry (GC-MS) in selected ion monitoring and in full scan mode (mass range, 50–550 amu). N-Phenylcarbazole was added into the nitrogen compound enriched fraction prior to GC-MS analysis.

For the analysis of C_0–C_3–phenols, a modification of the method described in Bennett et al. (1996) was employed. An aliquot (ca. 100 mg) of extract was loaded on to a C18 NEC solid phase extraction cartridge. Phenol-d6 (to check for potential loss of volatile phenol) and 2,4-dimethylphenol-d3 were also added as internal standards. The hydrocarbons (as above) were removed in n-hexane while the C_0–C_3–phenols were retained by the C18 NEC phase. The C_0–C_3–phenols are removed from the C18 NEC phase by eluting 5 ml of CH_2Cl_2. The solvent volume is gently reduced (taking care not to lose volatile phenol and cresols), to a minimum of about 200 µl. An aliquot of 2-naphthol was added as a check for recovery of the 2,4-dimethylphenol-d3. The C_0–C_3–phenols plus standards containing isolate was treated with bis-silyltrimethylfluoroacetamide (BSTFA) in preparation for analysis by GC-MS.

Gas Chromatography-mass Spectrometry (GC-MS)

Mass spectral characterisation of compounds in the various fractions was carried out using GC-MS on a Hewlett-Packard 5890 GC (using splitless injection) interfaced to a HP 5970B quadrupole mass selective detector (electron energy 70 eV, source temperature 250 °C). The saturated hydrocarbons, aromatic hydrocarbons and polar non-hydrocarbon fractions were analysed on a fused silica capillary column (HP-5, 95/5% methyl/phenyl silicone; dimensions 30 m × 0.32 mm i.d. × 0.25 µm film thickness). The GC oven temperature programs were as follows: aromatic hydrocarbons and C0–C3–phenols held at 40 °C for 2 min then programmed at 4 °C min− 1 to 300 °C and held at the final temperature for 20 min; saturated hydrocarbons and nitrogen compounds held at 40 °C for 2 min then programmed at 10°C min− 1 to 150°C and then 4 °C min− 1 to 300 °C and held at the final temperature for 20 min.

Compound identification was based on relative retention times, comparison of mass spectra with published mass spectra and in some case by co-chromatography with authentic standards.

Environmental Scanning Electron Microscopy

Studies of the core chips were carried out using a Philips/FEI XL30 Environmental SEM equipped with a LaB6 gun, and an EDAX UTW energy dispersive X-ray detector. Thin chips of each sample were prepared (freshly broken from core samples), each in the order of 10 mm diameter and approximately 2 mm thickness. Thin samples were used to ensure efficient cooling and maintenance of the correct sample surface temperature by the Peltier stage. Pre-cooled samples were placed onto the Peltier stage, which was maintained at 5 °C. H2O mode was selected from the ESEM controller, which utilises water vapour as the imaging gas. The chamber was then pumped to 5 Torr, and automatically passed through 5 pressure cycles from 4 to 10 Torr. The latter provides a water vapour atmosphere suitable for charge suppression and image amplification, as well as providing the source of wetting agent (water). Chamber pressure was then increased to 6.5 Torr, with the aim of condensing water droplets on to the surface of the sample as relative humidity approaches 100%. Observations on water droplet morphology can then be made. This method is essentially that followed by Wei et al. (2002) in a study of the wettability of fibres.

RESULTS AND DISCUSSION

Geochemical Proxies for Wettability Description

We have been studying the partition behaviour of polar petroleum compounds in petroleum systems for several years in order to produce estimates of the relative volumes of oil/water and rock interacting within such systems (Larter and Aplin, 1995). The groups of compounds that have received our attention include the aromatic oxygen (alkylphenols and alkylfluorenones) and aromatic nitrogen (alkylcarbazoles and benzocarbazole) compounds. Adsorption of organic species from petroleum to solid surfaces is widely discussed as a control on reservoir surface state and wettability. Adsorption of non-hydrocarbon species, such as asphaltenes, resins, or specific nitrogen or oxygen compounds, onto dry mineral surfaces directly from oil or solvent solution is easily demonstrated in the laboratory (Crocker and Marchin, 1988, Li et al., 1994, Li et al., 1995 and Taylor, 1994) but is not the common setting with initially water saturated reservoirs. With water wetted surfaces, the adsorption rates are dramatically reduced and with species that are insoluble in water, access to mineral surfaces through water films presents a major problem.

The partition behaviour of alkylphenols in crude oil: brine systems were investigated under reservoir conditions (Bennett and Larter, 1997 and Bennett et al., 2003). The similarity of alkylphenol distributions in brines from the laboratory simulation experiments and natural formation waters (North Sea) suggests partitioning of alkylphenols in crude oil: brine systems could explain the relative oil distributions, and may be an important process controlling the subsurface mobilisation of polar compounds. In addition to this evidence for rapid equilibration of alkylphenols with waters, flowing synthetic oils containing alkylphenols through rock slabs of sandstones resulted in rapid wettability changes (Huang et al., 1996), and this correlated with removal of alkylphenols from the synthetic oils. Thus short term wettability variations in petroleum systems may be facilitated by the rapid partitioning of hydrophilic compounds, such as alkylphenols, between the crude oil/brine/rock phases.

The behaviour of petroleum components following simulated petroleum migration through a continuous siltstone core (about 87 cm long × 5 cm wide) was investigated. Dramatic changes were observed in the composition and distributions of polar non-hydrocarbons compared to the original crude oil, namely in the fluoren-9-one, carbazole, benzocarbazole and C_0–C_3–phenols.

Bennett et al. (2004) showed that C_0–C_3phenols in produced oils only amounted to 5% recovery compared to the C_0–C_3–phenols in the original oil, while fluoren-9-one, carbazole and benzocarbazoles were ultimately recovered up to 90% of the original quantities. The core extract petroleum was also investigated to monitor compositional changes along the core (Bennett et al., 2004). With respect to distance from the inlet, fluoren-9-one concentrations were similar (Fig. 1a), while carbazole concentrations decreased sharply at the outlet (Fig. 1b), thus indicating propagation along the core was relatively unperturbed. Meanwhile, C_0–C_3phenol concentrations decreased from inlet to outlet depicting a slow moving front propagating through the core suggesting these species have a higher sensitivity for crude oil/brine/rock interactions (Fig. 1c). Thus, during the migration process the C_0–C_3phenols are effectively removed!.

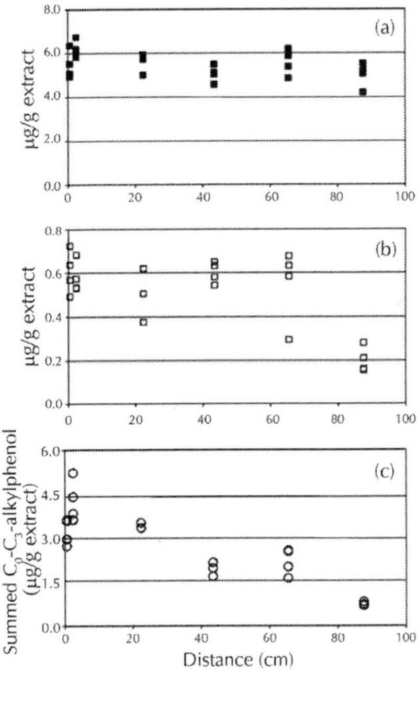

Figure 1: Plot showing change in concentrations (µg/g extract) of (a) fluoren-9-one, (b) carbazole and (c) total summed C_0–C_3–phenols

in core extract petroleum recovered from core samples as a function of distance (cm) along the core (modified from Bennett et al., 2004).

Environmental Scanning Electron Microscopy (ESEM) was used to observe changes in surface water morphology during a condensation/sublimation cycle of water upon freshly exposed core material. The morphology and mode by which water spreads across a surface during condensation/sublimation cycle can be used to give an indication of the hydrophilic/hydrophobic tendency of surfaces examined under ESEM conditions, thus indicating the nature of the surface wettability. During a condensation cycle, if water spreads evenly across the surface as a continuous film it is said to be displaying hydrophilic behaviour and the sample is considered water-wet. In contrast, if water appears as discrete globules on a surface then the observation suggests hydrophobic behaviour indicating a shift towards more oil-wet conditions.

Fig. 2a to c show the ESEM images of water spreading behaviour on core chips from different locations along the core. The behaviour of water condensation on a sample from the inlet end of the core displays water as discrete globules with high contact angles indicating hydrophobic surfaces (Fig. 2a). Samples prepared from both the original core and core material sectioned towards the outlet of the core-flood displayed surfaces exhibiting hydrophilic tendencies. The ESEM image of a core chip from the outlet end (and the original core material prior to oil flooding) of the core-flood show water spreading across the surface in a continuous sheet indicates hydrophilic behaviour (Fig. 2c). Fig. 2b, from the mid-point of the core shows water spreading phenomena intermediate to the samples described as showing hydrophilic and hydrophobic behaviour. Fig. 2d shows an attempt to describe the wetting tendencies of all core samples analysed by using ESEM observations. The observations support previous results describing the hydrophobic surfaces from the inlet part of the core-flood (Bennett et al., 2004). The analysis of 8 samples along the core indicates that there is a progressive change in wettability propagated along the core from hydrophobic to hydrophilic as we move from the inlet to the outlet ends of the core. Graue et al. (2002) have also observed wettability alteration during core-flood experiments using chalk cores.

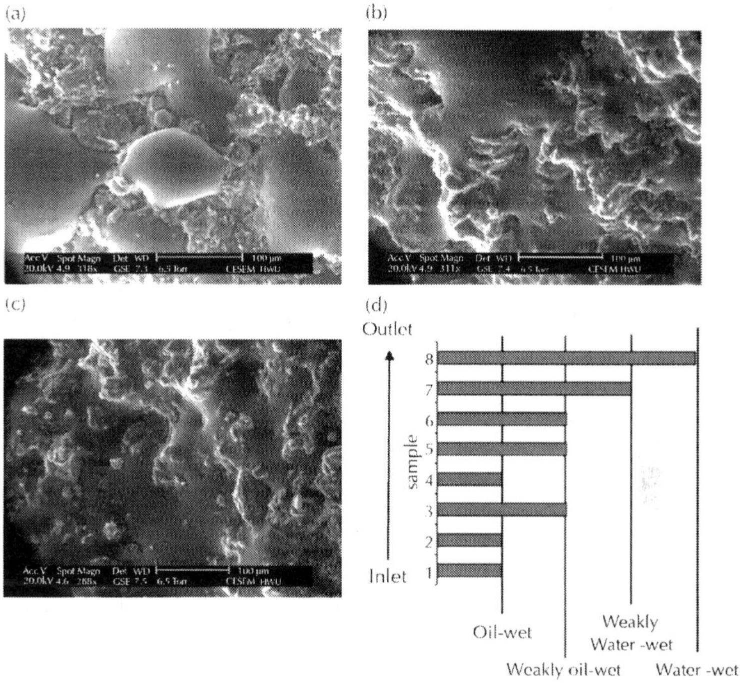

Figure 2: ESEM images depicting changes in relative hydrophilic/hydrophobic behaviour of water interactions on core chips from the core-flood experiment from the (a) inlet (sample 1; d), (b) middle (sample 3) and (c) at the outlet end (sample 8) of the core and (d) a qualitative description of wettability in samples along the core.

Wettability was most strongly altered at the core inlet suggesting surface active compounds are rapidly removed from the migrating petroleum. Bennett et al. (2004) showed the concentrations of C_0–C_3–phenols in core extracts decreased gradually along the core. Thus the ESEM descriptions of the interaction of water upon core surfaces appear to be consistent with the compositional changes encountered in the core extract petroleum which show the removal of low molecular weight hydrophilic components including alkylphenols.

Bennett et al. (2004) utilised results from model compound experiments and the molecular modelling approach employed by van Duin and Larter (2001), van Duin et al. (2002) to develop a hypothesis to understand the processes that may influence wettability alteration. The alkylphenols represent ideal compounds to investigate processes at the pore level since they are water soluble and are able to partition between all phases in the crude oil/brine/rock system. Indeed, compounds that are insoluble or weakly soluble in water are unlikely to directly access mineral surfaces that are surrounded by a water film. In order to gain access to the mineral surfaces nonpolar compounds require a surface preconditioning mechanism to disrupt the water film. The breakdown of the water film may be facilitated by alkylphenols and related compounds that are able to access mineral surfaces by partitioning into the water film on mineral surfaces rendering the mineral surface more hydrophobic (oleophilic) and effectively allowing high molecular weight compounds to then interact with the mineral surfaces.

In this study we have focused on a small component of petroleum surface active compounds. Further work might investigate a larger suite of compounds that is available to advanced techniques such as high resolution mass spectrometry. Marshall and Rodgers (2004) have proposed ion cyclotron resonance for determination of multiple polar NSO compound types by a single sample analysis to produce compositional map, which they term petroleomics. Hughey et al. (2004) demonstrated the application of electrospray high field Fourier transform ion cyclotron resonance mass spectrometry to the determination of acidic and neutral polar NSO compounds in Smackover oils of different thermal maturity. A preliminary study using High Resolution Mass Spectrometry of oil samples from the core flood also suggested that removal of low molecular weight multi-functional (containing one or more nitrogen, oxygen and or sulphur) compounds including aromatic oxygen compounds such as phenols were removed from the migrating petroleum (Larter et al., 2002).

The results show that the C_0–C_3–phenols display a higher degree of surface activity during the migration experiment compared to alkylfluorenones and alkylcarbazoles as they are not recovered from the core by continued oil flow. An equivalent comparative study of polar non-hydrocarbon composition of natural petroleum system sample sets are lacking though we know that alkylphenols and even carbazoles are removed during petroleum migration. The behaviour of C_0–C_3–phenols is now widely known in petroleum systems (Ioppolo et al., 1992 and Taylor et al., 1997). The total concentrations of C_0–C_3–phenols show a regular decrease with increasing migration distances in the Tampen Spur petroleum system (e.g. Bennett et al., 1996) over distances of a few tens of kilometers. Larter et al. (1996) show that the concentrations of benzocarbazoles decrease systematically in petroleum with increasing petroleum secondary migration distances. In the Dhahaban petroleum system of Oman, Lucach et al. (2003) showed a significant decrease in total C_0–C_3–phenols over the first 50 km within a petroleum migration system extending to 260 km, whereas benzocarbazole data was reported in all oils analysed up to the 260 km migration route (Terken and Frewin, 2000). In summary, polar nonhydrocarbons interact with rocks even with water films present and in the case of low molecular weight compounds such as alkylphenols appear able to change wettability on human timescales (i.e. core-flood experiment).

Linkages between Geochemical Proxies and Petrophysics of Reservoir Rocks

There are few reports in the literature describing integrated studies of petrophysical data with reservoir geochemical data, for example Larter et al. (1997). Our research reveals that several compounds residing in the polar non-hydrocarbon fraction of petroleum correlate with petrophysical/core properties (Fig. 3, Fig. 4 and Fig. 5). Fig. 3 shows a plot of the fluoren-9-one concentrations in core extract petroleum, along with the wireline gamma ray log for the same 24 ft interval for sample set-1. The strongest deflections in both the gamma ray log and fluoren-9-one chemical log are observed in the basal coarsening upwards unit. The response in the gamma ray log is due to the increasing clay content towards the base of the core, thus it appears that fluoren-9-one concentration may also be sensitive to clay content. Strong correlations between other geochemical parameters and the gamma ray log, particularly in the basal coarsening upwards unit, are also shown by the C_2–phenols (Fig. 4) and the benzocarbazoles (Fig. 5). A dramatic change in the ratio of shielded (methyl groups adjacent to the hydroxyl functionality) and exposed (methyl groups located away from the hydroxyl functionality) isomers in the C_2–phenols also corresponds to the basal coarsening upwards sequence. The result appears counter-intuitive since it might be expected that the exposed isomer, 3,5-dimethylphenol (3,5-DMP), would adsorb onto clays more readily, while the shielded isomer, 2,6-dimethylphenol (2,6-DMP), is hindered from such interactions. Fig. 5 shows that aromatic nitrogen compounds (benzo[*a*]carbazole and benzo[*c*]carbazole isomers) also display the highest response towards the base of the coarsening upwards sequence. Irrespective of the reason for the correlations the results demonstrate that petrophysical logs are tracking rock properties that are in some way responsible for the relative distribution of petroleum components in the crude oil/brine/rock system.

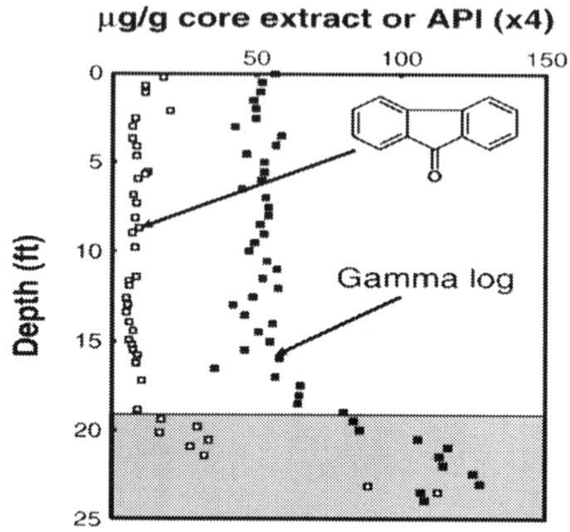

Figure 3: Comparison of gamma ray log (API × 4) and fluorenone concentrations (μg/g extract) versus depth in core extracts from a North Sea reservoir (sample set-1).

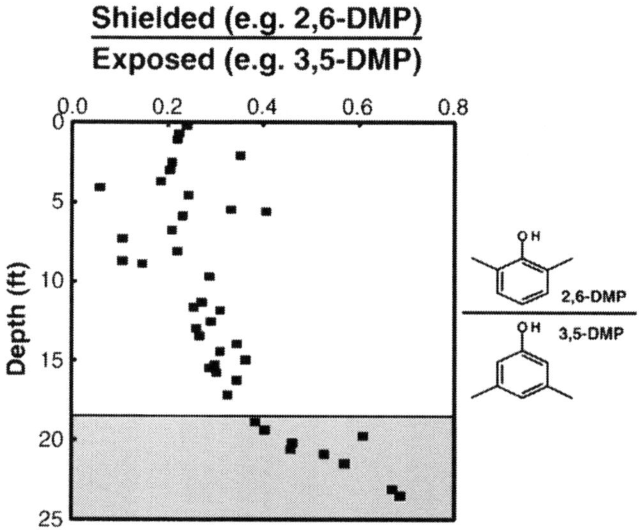

Figure 4: Measure of the behaviour of 2, 6-dimethylphenol/3,5-dimethylphenol with depth in reservoir cores (sample set-1).

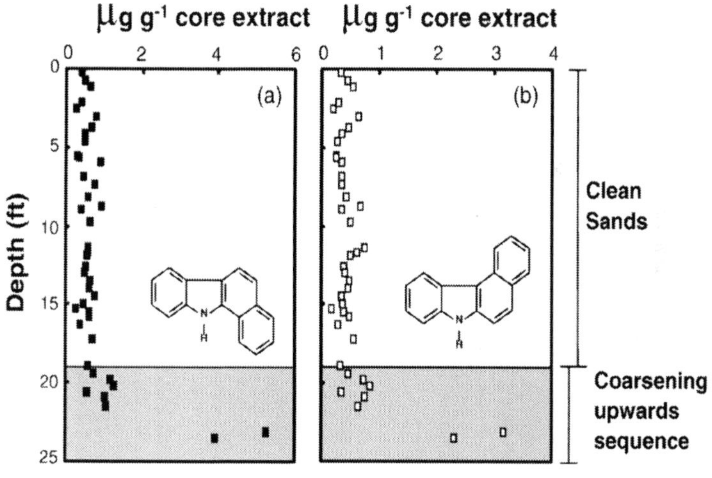

Figure 5: Plots of the concentration (μg/g extract) of (a) benzo[*a*] carbazole and (b) (benzo[c]carbazole versus depth in core extracts from North Sea reservoir cores (sample set-1).

Further studies of other sedimentary genetic units (sample set-2) also show strong variations in geochemical parameters correlating with the gamma ray log. In Fig. 6a, the total gamma ray log, reveals that the section of interest consists of two major lithologies. From 10 to 26 ft depth the lithology comprises a clean sand unit, while the lower section (26 to 36 ft) consists of fine grained micaceous sandstones. The higher gamma ray signal in this lower section is due to the increased mica content. Fig. 6b shows a chemical log derived from the concentration of fluoren-9-one in the core extract petroleum, which can be compared with the gamma ray log (Fig. 6a). A strong correlation also exists between the ratio 1-methylcarbazole/4-methylcarbazole (Fig. 6c) and the total gamma ray log emphasising that petroleum components appear to be tracking petrophysical properties.

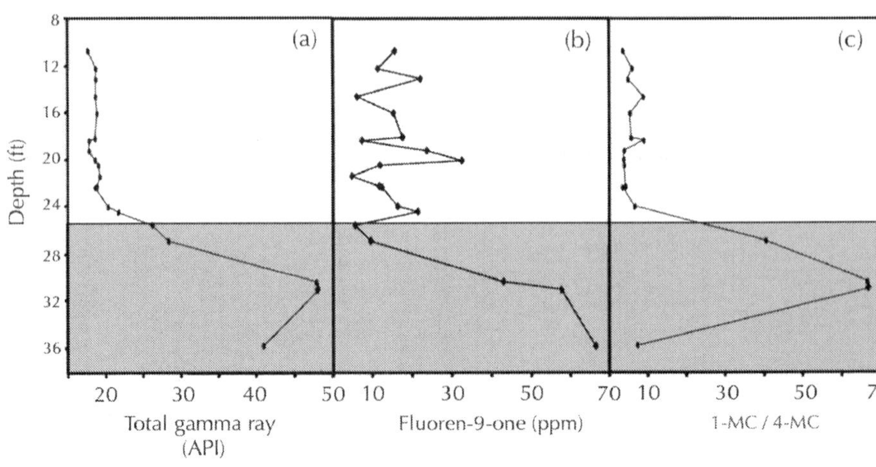

Figure 6: Comparison between (a) total gamma ray, (b) fluorenone concentrations (ppm or μg/g extract) and (c) 1-methylcarbazole/4-methylcarbazole (1-MC/4-MC) relative to depth (ft) in North Sea reservoir sample set-2.

Fig. 7a shows a plot of the U/Th ratio, inferred from the spectral gamma signal, and displays three perturbations at 12 to 14 ft, 17 to 19 ft and 22 to 25 ft. Differences in the bedding parallel and bedding perpendicular acoustic measurements, along with X-ray images suggested these Uranium enriched bands were laminated sands. X-ray Diffraction (XRD) analysis of the core samples indicated that the laminated sands contain mainly muscovite mica as the nonquartz fraction. Fig. 7b and c show that some geochemical parameters such as the ratio 4-methylfluorene/4-methylfluoren-9-one and the ratio 3-methylfluorene + 2-methylfluorene/fluorene + 1-methylfluorene + 4-methylfluorene (MFI) exhibit increases at the same intervals where the U/Th ratio increases. However, in the lower sandstone at 31 to 32 ft, where the U/Th ratio is relatively low, both geochemical parameters increase. This increase found in the geochemical parameters in the lower zone, coincides with an illite rich interval, thus in general the geochemical ratios appear to be responding to increasing clay content in the sandstones.

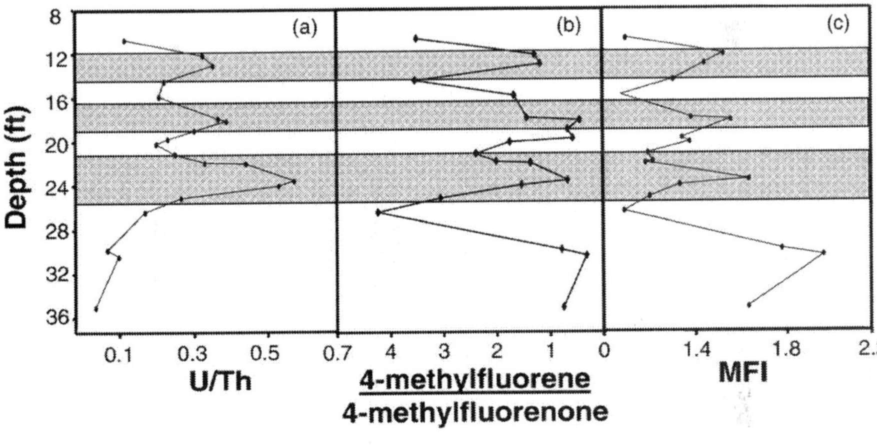

Figure 7: Comparison between (a) uranium/thorium (U/Th) spectral gamma ray log, (b) 4-methylfluorene/4-methylfluorenone and (c) MFI (see text for description) versus depth (ft) in North Sea reservoir sample set-2.

Sample sets 3 and 4 consisting of clastic reservoir cores were extracted using organic solvents to remove the petroleum and then were submitted to X-ray diffraction analysis. The XRD measurements were used to define the relative clay versus quartz content utilising the reflections typical of clay (19.9° 2θ) and quartz (26.6° 2θ).

The relative clay content was plotted against the concentrations of benzo[*a*]carbazole and benzo[*c*]carbazole in the core extract petroleum from these samples (Fig. 8). The relationships show a great deal of scatter, but reveal on the whole higher benzocarbazole concentrations with increasing clay content. Clay minerals are implicated in the removal of organic nitrogen compounds from petroleum (Charlesworth, 1986). Model experiments flowing characterised oils through initially water saturated sandpacks containing added kerogen or clay minerals, both resulted in reduction of the benzo[*a*]carbazole/benzo[*c*]carbazole ratio (Chen, 1995) in the oils. We also found correlations amongst the fluorenone compounds. Fig. 9 shows the XRD based clay content versus the fluoren-9-one/methylfluorenone ratios.

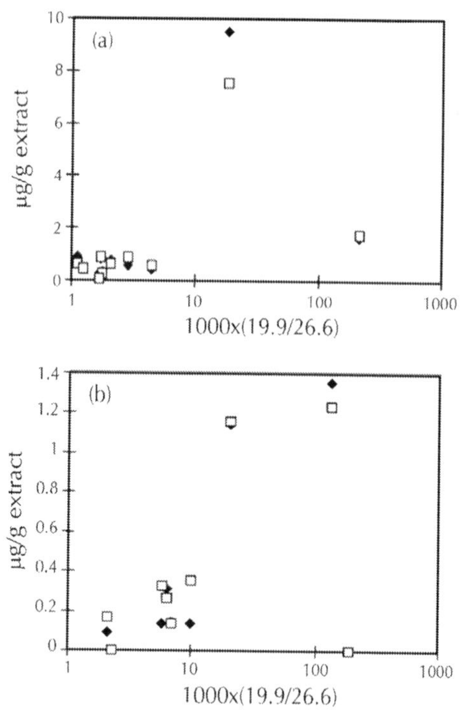

Figure 8: Plots of relative clay content defined by the ratio of the clay (19.9° 2θ) to quartz (26.6° 2θ) reflections × 1000 versus benzo[*a*]

carbazole (diamonds) and benzo[c]carbazole (squares) concentrations (μg/g extract) for core samples obtained from a Palaeocene reservoir of the North Sea (a) sample set-3, and (b) sample set-4.

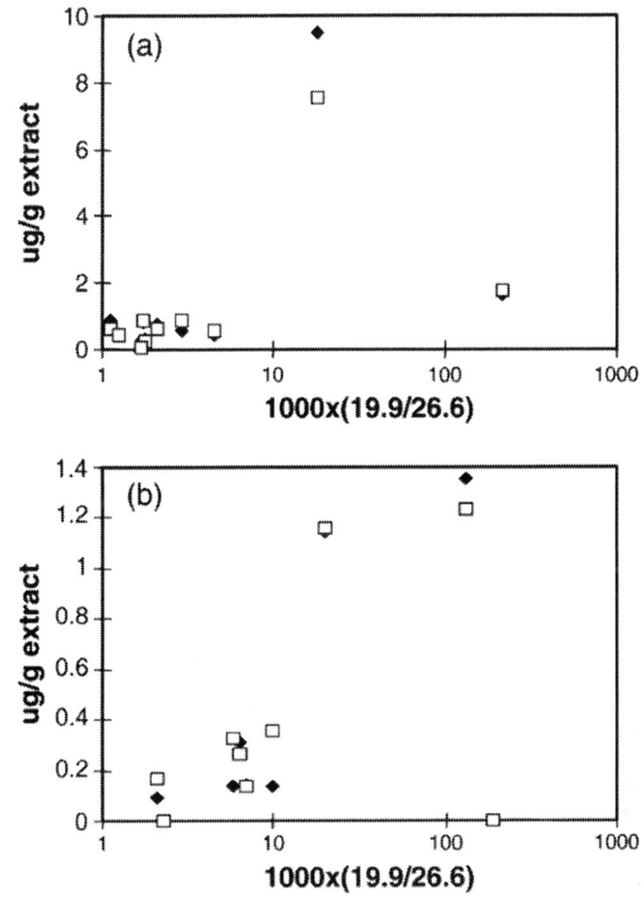

Figure 9: Plots of fluorenone/methylfluorenone ratio versus relative clay content defined by the ratio of the clay (19.9° 2θ) to quartz (26.6° 2θ) × 1000 for core samples obtained from a Palaeocene reservoir of the North Sea (a) sample set-3, and (b) sample set-4.

In summary, a combination of geochemical and petrophysical measurements of petroleum sandstone reservoirs showed that geochemical logs based on alkylfluorenones, C_0–C_3–phenols and benzocarbazoles (aromatic nitrogen compounds) correlated well with clay content measured by gamma ray logs suggesting that these non-hydrocarbons were selectively concentrated in certain reservoir lithologies. The geochemical data may be compiled into chemical logs for comparison with petrophysical and core data to allow identification of key zones where fluid behaviour may be different with regard to wetting behaviour.

CONCLUSIONS

Polar non-hydrocarbon compounds are sensitive to chemical and physical interactions in the crude oil/brine/rock system. ESEM showed, on a timescale of weeks, surface wetting tendencies for oil in contact with initially water-wet core changes from hydrophilic to hydrophobic. Wettability changes described by ESEM appeared to correspond to changes in hydrophilic non-hydrocarbon compounds such as the C0–C3–phenols, though multifunctional compounds with one or more N, S or O atom are likely to be important facilitators of wettability alteration.

Correlations exist between molecular markers, such as carbazole, fluorene and fluoren-9-one and total and spectral gamma ray. It was shown that the chemical composition of oil trapped in pores is influenced by several factors including the amount and type of clay minerals. These observations could be used for predicting petrophysical properties. Molecular parameters based on the polar non-hydrocarbon composition of petroleum could be used to differentiate between different units, such as clean sand units and adjacent coarsening upward muddy sand sequences, showing that reservoir geochemical parameters provide an independent response to those properties defined by petrophysical methods. While these correlations exist it is not always possible to tell if the compositional changes occurred in the subsurface or after sampling and storage. Nonetheless compositional change does seem to reflect mineralogy and rock environment.

ACKNOWLEDGEMENTS

The core-flood experiment was carried out under the EU FP4 SMACCERS research program. The sponsors of PEGASUS Phase I (EPSRC, DTI, Amerada Hess, Amoco, British Gas, BHP Petroleum, Chevron UK, Fina, Saga Petroleum, Norsk Hydro, Schlumberger, Shell UK, Statoil and Texaco) and PEGASUS Phase II (EPSRC, DTI, Amerada Hess, BG, Halliburton and Schlumberger) are gratefully acknowledged for encouragement and support. Professor Patrick Corbett is acknowledged for useful discussions. Kim Noke and Paul Donohoe are thanked for GC-MS assistance. Additional support from NSERC, AIF and the Canada Research Chair program are also acknowledged (Larter, Bennett). We would also like to thank Drs. Craig Smalley and Luiz Trindade for their constructive criticisms and comments made on an earlier version of this manuscript.

REFERENCES

1. Akhlaq, M.S., Kessel, D., Dornow, W., 1996. Proceedings 3rd International Symposium on Evaluation of Reservoir Wettability and its Effect on Oil Recovery, pp. 63–67.
2. Anderson, W.G., 1986. Wettability literature survey: Part 1 — rock/oil/ brine interactions and the effects of core handling on wettability. J. Pet. Technol. 1125.
3. Bennett, B., Larter, S.R., 1997. Partition behaviour of alklphenols in crude oil/brine systems under subsurface conditions. Geochim. Cosmochim. Acta 61, 4393–4402.
4. Bennett, B., Larter, S.R., 1998. The 5th International Symposium on Evaluation of Reservoir Wettability and its Effect on Oil Recovery, Trondheim, Norway, 22–24 June, pp. 70–79.
5. Bennett, B., Larter, S.R., 2000. Quantitative separation of aliphatic and aromatic hydrocarbons using silver ion-silica solid-phase extraction. Anal. Chem. 72, 1039–1044.

6. Bennett, B., Bowler, B.F.J., Larter, S.R., 1996. Determination of C0–C3 alkylphenols in crude oils and waters. Anal. Chem. 68, 3697–3702.
7. Bennett, B., Chen, M., Brincat, D., Gelin, F.J.P., Larter, S.R., 2002. Fractionation of benzocarbazoles between source rocks and petroleums. Org. Geochem. 33, 545–559. Bennett, B., Aplin, A.C., Larter, S.R., 2003. Measurement of partition coefficients of phenol and cresols in gas charged crude oil/water systems. Org. Geochem. 34, 1581–1590.
8. Bennett, B., Buckman, J.O., Bowler, B.F.J., Larter, S.R., 2004. Wettability alteration in petroleum systems: the role of polar nonhydrocarbons. Pet. Geosci. 10, 271–277.
9. Bennett, B., Lager, A., Larter, S.R., 2007. Polar non-hydrocarbons in crude oil and rock extracts: Recovery and impact of sample storage protocols. J. Pet. Sci. Eng. 58, 367–374, doi:10.1016/j.petrol.2006.06.008.
10. Buckley, J.S., Liu, Y., Xie, X., Morrow, N.R., 1997. Asphaltenes and crude oil wetting — the effects of oil composition. J. Soc. Pet. Eng. 2, 107–119.
11. Buckley, J.S., Liu, Y., Monsterleet, S., 1998. Mechanisms of wetting alteration by crude oil. J. Soc. Pet. Eng. 3, 54–61.
12. Charlesworth, J.M., 1986. Interaction of clay minerals with organic nitrogen compounds released by kerogen pyrolysis. Geochim. Cosmochim. Acta 50, 1431–1435.
13. Chen, M., 1995. Response of pyrrolic and phenolic compounds to petroleum migration and in-reservoir processes. PhD thesis, University of Newcastle upon Tyne.
14. Crocker, M.E., Marchin, L.M., 1988. Wettability and adsorption characteristics of crude-oil asphaltene and polar fractions. J. Pet. Technol. 470–474.
15. Cuiec, L., 1977. Study of the problems related to the restoration of the natural state of core samples. J. Can. Pet. Technol. 16, 470–474.

16. Cuiec, L., 1984. Rock/crude oil interactions and wettability: an attempt to understand their interrelation. SPE AIMME 13211.
17. England, W.A., 1990. The geochemistry of petroleum reservoirs. Org. Geochem. 16, 415–425.
18. Graue, A., Aspnenes, E., Bogno, T., Moe, R.W., Ramsdal, J., 2002. Alteration of wettability and wettability heterogeneity. J. Pet. Sci. Eng. 33, 3–17.
19. Huang, Y., Ringrose, P.S., Sorbie, K.S., Larter, S.R., 1996. The effects of heterogeneity and wettability on oil recovery form laminated sedimentary structures. SPE J. 1, 451–461.
20. Hughey, C.A., Rodgers, R.P., Marshall, A.G., Walters, C.C., Qian, K.N., Mankiewicz, P., 2004. Acidic and neutral polar NSO compounds in Smackover oils of different thermal maturity revealed by electrospray high field Fourier transform ion cyclotron resonance mass spectrometry. Org. Geochem. 35, 863–880.
21. Ioppolo, M., Alexander, R., Kagi, R.I., 1992. Identification and analysis of C0–C3 phenols in some Australian crude oils. Org. Geochem. 18, 603–609. Kaminsky, R., Radke, C.J., 1997. Asphaltenes, water films, and wettability reversal. SPE J. 485–493.
22. Larter, S.R., Aplin, A.C., 1995. The geochemistry of reservoirs. Spec. Publ. - Geol. Soc. Lond. 86, 5–32.
23. Larter, S.R., Bowler, B.F.J., Li, M., Chen, M., Brincat, D., Bennett, B., Noke, K., Donohoe, P., Simons, D., Kohnen, M., Allan, J., Telnaes, N., Horstad, I., 1996. Molecular indicators of secondary oil migration distances. Nature 383, 593–597.
24. Larter, S.R., Aplin, A.C., Corbett, P.W.M., Ementon, N., Chen, M., Taylor, P.N., 1997. Reservoir geochemistry: a link between reservoir geology and engineering? SPE Res. Eng. 12, 12–17.
25. Larter, S.R., Bowler, B.F.J., Clarke, E., Wilson, C., Moffatt, B., Bennett, B., Yardley, G., Carruthers, D., 2000. An experimental investigation of geochromatography during secondary

migration of petroleum performed under subsurface conditions with a real rock. Geochem. Trans. vol. 9.

26. Larter, S.R., Bennett, B., van Duin, A.C.T., Bowler, B.F.J., Bond, K., Aplin, A.C., Buckman, J.O., Winans, R., di Primio, R., 2002. Capillary seal or permeability barrier: field, experimental and theoretical considerations of caprock wetting state. AAPG Hedberg Conference Evaluating the Hydrocarbon Sealing Potential of Faults and Caprocks, December 2–5, Barossa Valley, South Australia.
27. Li, M., Larter, S.R., Frolov, Y.B., 1994. Adsorptive interaction between petroleum nitrogen compounds and organic and/or B. Bennett et al. / Journal of Petroleum Science and Engineering 58 (2007) 355–366 365 mineral phases in subsurface rocks: models for compositional fractionation of pyrrolic nitrogen compounds in petroleum during petroleum migration. J. High Resolut. Chromatogr. 17, 230–236.
28. Li, M., Larter, S.R., Stoddart, D.P., Bjøroy, M., 1995. Fractionation of pyrrolic nitrogen compounds in petroleum during reservoir filling: derivation of migration related geochemical parameters. In: England, W.A., Cubitt, J. (Eds.), Reservoir Geochemistry, Geological Society, Special Publication.
29. Lucach, S.O., Bowler, B.F.J., Frewin, N., Larter, S.R., 2003. Variation in alkylphenol distributions in a homogenous oil suite from the Dhahaban petroleum system of Oman. Org. Geochem. 33, 581–594.
30. Marshall, A.G., Rodgers, R.P., 2004. Petroleomics: the next grand challenge for chemical analysis. Acc. Chem. Res. 37, 53–59.
31. Mitchell, A.G., Hazell, L.B., Webb, K.J., 1990. Wettability determination: pore surface analysis. SPE AIMME 2050, 351–360. Morrow, N.R., 1990. Wettability and its effect on oil recovery. J. Pet. Technol. 42, 1476–1484.
32. Stainforth, J.G., 2004. New insights into reservoir filling and mixing processes. In: Cubitt, J.M., England, W.A., Larter,

S.R. (Eds.), Understanding Petroleum Reservoirs: Towards an Integrated Engineering and Geochemical Approach. Geological Society, London, Special Publications, vol. 237, pp. 115–132.

33. Stoddart, D.P., Hall, P.B., Larter, S.R., Brasher, J., Li, M., Bjøroy, M., 1995. The reservoir geochemistry of the Eldfisk Field, Norwegian North Sea. In: Cubitt, J.M., England, W.A. (Eds.), the Geochemistry of Reservoirs. Geological Society, London, Special Publication, vol. 86, pp. 257–279.
34. Taylor, P.N., 1994. The petroleum geochemistry of phenols. PhD thesis, University of Newcastle upon Tyne.
35. Taylor, P.N., Larter, S.R., Jones, D.M., Dale, J.D., Horstad, I., 1997. Geochim. Cosmochim. Acta 61, 1899–1910.
36. Terken, J.M.J., Frewin, N.L., 2000. Dhahaban petroleum system of Oman. AAPG Bull. 84, 523–544.
37. Treiber, L.E., Archer, D.L., Owens, W.W., 1972. A laboratory evaluation of the wettability of fifty oil-producing reservoirs. J. SPE 12, 531–540.
38. Van Duin, A.C.T., Larter, S.R., 2001. Molecular dynamics investigation into the adsorption of organic compounds on kaolinite surfaces. Org. Geochem. 32, 143–150.
39. Van Duin, A.C.T., Bennett, B., Larter, S.R., 2002. Computational chemical study on the displacement of water films near mineral surfaces. 7th International Symposium on Reservoir Wettability, Freycinet, Tasmania, Australia 12–15 March.
40. Wei, Q., Mather, R.R., Fotheringham, A.F., Yang, R.D., 2002. Observation of wetting behaviour of polypropylene microfibres by environmental scanning electron microscopy. J. Aerosol Sci. 33, 1589–1593.

Chapter 9

Evaluation and Comparison of Enrichment Efficiency of Physical/Chemical Activations and Functionalized Activated Carbons Derived from Fluid Petroleum Coke for Environmental Applications

N. Rambabu[a], R. Azargoha[a], A.K. Dalai[a], and J. Adjaye[b]

[a]Catalysis and Chemical Reaction Engineering Laboratories, Department of Chemical and Biological Engineering, University of Saskatchewan, Saskatoon, SK, Canada S7N 5A9

[b]Syncrude Edmonton Research Centre, Edmonton, AB, Canada T6N 1H4

ABSTRACT

Activated carbons were produced from fluid petroleum coke using physical (steam and CO_2) and chemical (KOH and H_3PO_4) activations. Effects of process parameters including activation temperature, activation time period and water flow rate on yield and BET surface area of steam-activated carbons were investigated and optimized. Steam activation improved BET surface area of the precursor more than 40 times at optimum operating conditions and KOH activation increased BET surface area from 11 to 692 m^2/g. Properties of activated carbons were characterized by BET, SEM, CHSNO, FTIR, ICPMS and Boehm analysis techniques. The steam-activated carbons were treated with HNO_3 and H_2O_2 separately to study the effects of acid treatments on surface chemistry, porous characteristics and environmental applications of activated carbons. Activated carbons produced in this study were used as adsorbent for ammonium ion (40.0 and 260.0 mg/L) in liquid phase and as catalyst for direct oxidation of H_2S (10,000 ppm) in gas stream. Among all other samples HNO_3 and H_2O_2-treated samples showed the highest ammonium adsorption by 184.6 and 169.6 mg/g, respectively. The largest breakthrough time (320 min) among all samples was shown by KOH activated carbon. The SO_2 production was less than 5% of H_2S fed to the reactor.

INTRODUCTION

Petroleum coke including delayed coke, fluid coke, needle coke, shot coke, and flexi coke is a by-product of the upgrading process for oil sand derived bitumen [1]. Petroleum coke is used as fuel in refineries and as raw material in cement, steel, and aluminum industries [2]. Several studies have looked at ways to utilize the petroleum coke for gasification [3], water purification [4], steam and electricity generation [1] and [5], and as a fuel in chemical-looping combustion processes [6]. Due to its large carbon content (> 80 wt.%) and relatively low ash content, this petroleum coke can be converted to a value-added product such as activated carbon, which may have many catalytic and adsorptive applications.

Activated carbons have been produced from different types of petroleum coke. For example, Lee et al. (2000) produced activated carbon from delayed petroleum coke obtained from Hyundai (HD) oil refinery with a BET surface area of 1980 m^2/g at alkali to coke in the ratio of 1 to 4 [7] and Lu et al. (2010) prepared activated carbon with the BET surface area of 1763 m^2/g with KOH/coke ratio of 3 [8]. Wu et al. (2005) produced porous carbons with a surface area range of 2500–3000 m^2/g from Shengli petroleum coke (SPC) and Minxi petroleum coke (MPC) by different activation methods such as H_2O, KOH and/or KOH + H_2O as active agents using KOH and H_2O co-activation with a weight ratio of KOH to petroleum coke of 2 to 1, in 30 min activation time at 800 °C [9]. Kawano et al. (2008) studied the different weight ratios of KOH to granular petroleum coke and observed a maximum BET surface area of 1798 m^2/g for 5 wt.% KOH at 800 °C in 2 h of activation time [10]. Otowa et al. (1997) produced high surface area (over 3000 m^2/g) activated carbon using KOH activation from MAXSORB petroleum coke with 4 wt.% KOH at 700 °C for 1 h [4]. Jiang et al. (2008) produced activated carbon from a typical Chinese petroleum coke provided by CNPC Liaohe Petrochemical Company with a surface area of 2408 m^2/g at experimental conditions of KOH at a mass ratio of 1:3, temperature of 1073 K and activation time of 1 h [11]. All of the above studies were for delayed coke, needle coke, shot coke, and flexi coke. Little information is available on the production of activated carbon from fluid coke [9] and [12] and showed that fluid coke is less reactive compared to other petroleum cokes, coals, and other carbonaceous materials due to combustion of its origin and production process [13].

The global production of petroleum coke was 80.815 MMTPA in 2000, and it was increased to 123.058 MMTPA in 2010. In 2015, the production is expected to grow to 161.271 MMTPA [14].

The coke produced by Syncrude is referred to as fluid coke and the technology used for primary upgrading is known as the fluid-coking process that can be seen from the literature [15] and [16]. Some studies have focused on the utilization of fluid coke for hydro desulphurization of athabasca fluid coke and oil sand cokes[13] and [17]. The adsorption efficiency and catalytic oxidation capacity of activated carbons are strongly influenced by the nature of the precursor, process parameters, pore size distribution, porous structure, and surface functionality [4] and [18]. The production of high surface area activated carbon from fluid coke is more difficult compared with that for delayed, flexi, and other cokes, due to the stable micrographitic structure and lack of initial pores of fluid coke (Syncrude petroleum coke).

The oil production in Canada in 2010 was 2.8 million barrels per day. The CAAP (Canadian Association of Petroleum Producers) estimate for 2025 is 4.7 million barrels per day and incorporates the most optimistic growth scenario for Canadian oil production [19]. Hydrotreating process is used for removing hetero-atoms such as sulfur, nitrogen, aromatics, metals, and other contaminants from oil feed stocks. In this process, some amounts of poisonous gases like H_2S and NH_3 are produced, which raises environmental concern. According to the U.S. Department of Health and Human Services and the Agency for Toxic Substances and Disease Registry (ATSDR), the toxicological profile for hydrogen sulfide in ambient air ranges from 0.11 to 0.33 ppb, but much higher concentrations, (> 90 ppb) have been detected in ambient air [20]. Ammonia exists naturally in the air at levels between 1 and 5 ppb. Due to their porous structure and flexible surface chemistry, activated carbons are known as effective adsorbents for many applications. They can capture positively charged ions like NH_4^+, K^+, Ca^{2+}, and Mg^{2+} that are retained on carbon surfaces and not lost through volatilization or leaching [21]. The largest energy input into crop production is generally nitrogen fertilizer. The binding of NH_4^+ to carbon surfaces is of particular interest because it reduces the rate of nitrification (NH_4^+ to NO_3^-), hence, the loss of N_2O and N_2 via denitrification, and it can sequester nutrients and hold water that might otherwise leach through the soil profile. Activated biochar is able to bind nitrogen so that less is lost to leaching and denitrification, leading to a reduced requirement for nitrogen fertilizer applications, as well as reduced ground water contamination and reduced greenhouse gas emissions (via nitrous oxide from the soil) [21]. For these reasons, more attention is required for the utilization of fluid coke, as well as methods for decreasing the H_2S and ammonium levels in ambient air and water respectively.

The objective of this research is to study the utilization of fluid coke for the production of large surface area activated carbons using physical (CO_2 and steam) and chemical (KOH and H_3PO_4) activations. Effects of post-treatments, such as HNO_3 and H_2O_2, on the porous structure and surface chemistry of steam-activated carbons were also studied. In addition, the process parameters for steam activation (water flow rate, activation time period and activation temperature) are optimized for the production of activated carbons with large yield and relatively high BET surface area. Evaluation and comparison of enrichment efficiency of oxidized and non-oxidized activated carbons for two different environmental applications, such as adsorption of ammonium ion from aqueous solution and direct catalytic oxidation of H_2S (1 mol% in CH_4) were also studied.

MATERIALS AND METHODS

Materials

Fluid petroleum coke was obtained from Syncrude Canada Ltd., Alberta, Canada and sieved to particles in a size range of 150 to 350 μm. The schematic diagram of the experimental set-up for physical and chemical activation processes is shown in Fig. 1. Specifications of the reactor for hydrogen sulfide oxidation are described in our earlier study [22]. Analytical grade chemicals (HNO_3, H_2O_2, H_3PO_4, KOH, and NH_4OH) were obtained from Alfa

Acer Chemicals. Nitrogen, air, SO_2 and H_2S gases were provided by Praxair Canada.

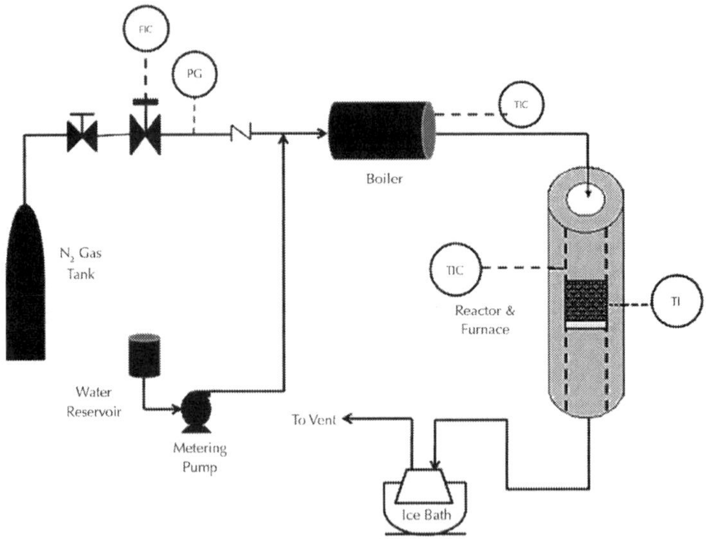

Figure 1: Schematic diagram of activation reactor setup.

Procedures for Chemical and Physical Activations

Samples were named according to their activation methods: raw petroleum coke (RPC), steam-activated carbon (SAC), steam activation + HNO_3 treated activated carbon (SAN), steam activation + H_2O_2 treated activated carbon (SAH), CO_2 activated carbon (CAC), phosphoric acid activated carbon (PAAC), and KOH activated carbon (KAC).

Chemical Activation

KAC and PAAC samples were prepared using a two-step process. For the first step, the chemical reagent 60 g (KOH or H_3PO_4) was dissolved in 100 ml of water and the solution was mixed with 20.0

g of raw petroleum coke (RPC). To ensure the access of chemical reagent to the interior of the petroleum coke, this mixture was shaken at 150 rpm for 2 h at room temperature (24 °C), and then dried at 120 °C in a hot air oven for overnight. 20.0 g of chemically impregnated coke was placed in the fixed-bed reactor under nitrogen flow (100 ml/min), which was heated up to 300 °C at heating rate of 4 °C/min, and held at this temperature for 1 h to prevent carbon loss through the direct attack of steam [4]. The temperature was then increased at the same heating rate to reach the desired activation temperature at 500 °C. This chemical activation time was chosen based on the similar work performed by Lee and Choi 2000 and DiPanfilo and Egiebor 1996 [7] and [12]. The products were then thoroughly washed with water, followed by treatment with 0.1 M HCl/0.1 M NaOH (depending on the type of chemical regent used for impregnation) to KOH and H_3PO_4, respectively, and finally washed with distilled water to remove the soluble salts and the potassium or phosphorous compounds. When the pH of the washed solutions was in the range of 7 ± 0.5, samples were dried at 110 °C for 12 h and characterized for physical and chemical properties.

Steam Activation

20.0 g of RPC was placed in the reactor for steam activation. The nitrogen flow was adjusted to 100 ml/min and the temperature was increased to the activation temperature at a heating rate of 4 °C/min. Temperature of the boiler was controlled by a temperature controller. After reaching the activation temperature of 900 °C, the injection of steam was started for the desired flow rate of 15 g/h and injection time of 9 h. After injection, the sample was cooled down under a slow ramp to reach the ambient temperature.

Acid Treatment

10.0 g of SAC sample was added to 64 ml of 70% HNO_3/ or 30% w/w H_2O_2 solution. This mixture was refluxed at 90 °C for 2.5 h (for both H_2O_2 and HNO_3) using a reflux column to recover the oxidizing reagent evaporated during the process. A magnetic stirrer was used for mixing during the reflux. The sample was washed with water to remove remaining acid until it reached the neutral pH. The product was dried overnight at 120 °C.

CO_2 Activation

20.0 g of RPC was loaded in the reactor situated in the furnace, and then heated at a heating rate of 4 °C/min under the N_2 gas flow rate of 100 ml/min. After reaching the activation temperature of 900 °C, the CO_2 gas was introduced at the flow rate of 200 ml/min for the desired activation time period of 9 h. After activation, solid samples were cooled to room temperature under the N_2 flow.

Characterization Techniques

The precursor and activated carbons were characterized using different characterization techniques such as BET surface area, scanning electron micrographs (SEM), elemental analysis (CHSNO), Boehm analysis, and Fourier transform infrared spectroscopy (FTIR) to study the physical and chemical characteristics. Boehm analysis was carried out to measure the concentration of acidic and basic groups on the carbon surface.

Porous Characteristics

BET surface area, total pore volume and average pore diameter of the precursor and activated carbons were determined by an automated gas adsorption analyzer, ASAP 2020 (Micromeritics, Instruments Inc., GA, USA). After degassing of samples at 300 °C to a vacuum of 550 μm Hg, nitrogen adsorption–desorption isotherms at – 196 °C were also measured by the ASAP 2020. The BET surface area was calculated using the BET equation [22]. For each analysis 0.2–0.3 g of sample was used. The accuracy of measurements performed by this equipment was ± 2%.

Elemental Analysis

The elemental analysis (CHNSO) of the petroleum coke as well as activated carbons was performed using Elementar Vario EL III.

FTIR Analysis

Chemical functional groups were examined by FT-IR using KBr pellet method at room temperature. The spectra were retrieved in the wave number range from 400 to 4000 cm^{-1} with a resolution of 4 cm^{-1} by an infrared spectrometer type PerkinElmer Spectrum 100 Series FTIR spectrometer (USA).

SEM Analysis

The morphological features of the samples were studied by electron micrographs using a JEOL 2011 SEM provided by the Department of Geology, University of Saskatchewan. The powder samples were placed on carbon tapes, and then coated with a thin layer of gold -palladium in an argon atmosphere using Agar Sputter Coater. The electron micrographs were recorded in a high vacuum mode under an accelerating voltage of 15 kV, using secondary electrons.

Boehm Titration

The concentrations of surface functional groups were measured using the Boehm titration method [23]. About 0.2 g of activated carbon was placed in 25 ml of each 0.05 N sodium hydroxide, 0.05 N sodium bicarbonate, 0.05 N sodium carbonate, and 0.05 N hydrochloric acid. The vials were shaken for 24 h and after filtration 5 ml of each filtrate was titrated with HCl or NaOH, depending on the original titrant used. The numbers of acidic functional groups of various types were calculated under the assumption that NaOH neutralizes carboxylic, phenolic, and lactonic groups; Na_2CO_3 neutralizes carboxylic and lactonic groups; and $NaHCO_3$ neutralizes only carboxylic groups. The number of surface basic sites was calculated from the amount of hydrochloric acid, which reacted with the carbon.

Ash Content

Ash content of all samples was determined according to ASTM D 2866-94.

ICP-MS

Analysis of the impurity composition (namely, Al, Ca, Cu, Fe, Li, Mg, Mn, P, Ti, V, and Zn) in the raw petroleum coke was performed using an inductively coupled plasma-mass spectrometer (ICP-MS, Model Sciex Elan 5000, manufactured by Perkin Elmer, USA).

Applications of Activated Carbons

Activated carbons were tested for environmental applications of ammonium ion adsorption from aqueous solutions and direct catalytic oxidation of H_2S from gas stream.

Procedure for Hydrogen Sulfide Oxidation

The reaction set-up and procedure for hydrogen sulfide oxidation were showed in our earlier studies [24]. 1.0 g of dry activated carbon sample was used for each H_2S oxidation experiment at the experimental conditions of pressure 700 kPa and temperature of 175 °C.

Procedure for Ammonium ion Adsorption

Ammonium ion adsorption experiments were conducted at two different ammonium hydroxide concentrations of 40.0 and 260.0 mg/L at 28 °C. 1.25 g of activated carbon was added to 125 ml of ammonium hydroxide solution. The sample bottles were placed in an orbital shaker at 200 rpm for equilibrium and adsorption studies. Blank runs (without an adsorbent) were performed with the same amount of ammonium hydroxide solution at the above operating conditions to serve as control experiment. After establishing equilibrium, the sample solutions were filtered to remove the carbon particles from the samples. These samples were used to determine the residual ammonium ion concentration in the solution using Colorimeter (MC-500) produced by Orbeco Hellige (Fl, USA) with *HACH* ammonium reagent bottles. The amount of adsorption *qe* (mg/g), at time t (h), was calculated by Eq. (1).

$$q_e = \frac{(C_0 - C_t)}{W} V \tag{1}$$

where *qe* is the amount (mg) of ammonium ion adsorbed per gram of adsorbent (mg/g), C_0 and *Ct* are, respectively, the initial and final concentrations of ammonium ion (mg/L), *V* is the volume of solution (L), and *W* is the weight of the adsorbent (g).

RESULTS AND DISCUSSION

Activated carbons were produced from Syncrude fluid petroleum coke using different activation methods. Six different types of activation/post-treatment methods, such as steam activation, steam activation + HNO_3 treatment, steam activation + H_2O_2 treatment, CO_2 activation, phosphoric acid activation and KOH activation were used to prepare the activated carbons from the Syncrude fluid petroleum coke. The effects of activation methods on the porous structure and the surface chemistry of activated carbons were investigated. The performance of activated carbons for ammonium ion adsorption from aqueous solution and direct catalytic oxidation of H_2S in gas stream were also studied. The porous characteristic analysis showed that RPC has a low surface area (11 m^2/g) and a very small micropore volume (0.001 cm^3/g), but these parameters are significantly improved with different activation methods which are described in following sections.

Steam Activation

Steam activation is a more cost-effective process compared with chemical activation. The possible reaction mechanism between steam and carbon is described by Muhlen and Van Heek (1995) [25]. Steam reacts with carbon and releases hydrogen and carbon monoxide, as shown in Eq. (2).

$$C+H_2O \rightarrow H_2+CO \quad (2)$$

At high temperatures, the water–gas shift reaction (Eq. (3)) and methanation reaction (Eq. (4)) can be catalyzed on the carbon surface and release the carbon dioxide, hydrogen as well as methane and water, respectively.

$$CO+H_2O \rightarrow CO_2+H_2 \quad (3)$$

$$CO+3H_2 \rightarrow CH_4+H_2O \quad (4)$$

Porous structure development is due to the penetration of oxidizing agents into the internal structure of the precursor and removal of unstable carbon atoms by reaction, which results in opening and widening of inaccessible pores [26]. This means that process parameters such as activation time, temperature and water flow rate play an important role in the development of the BET surface area and porous structure of activated carbons.

Effect of Activation Time

To study the effect of activation time on the BET surface area and porous structure of activated carbons produced from fluid coke, experiments were conducted at different reaction times ranging from 3 to 12 h, at water flow rate of 15.0 g/h and activation temperature of 900 °C. The results showed that an increase in the activation time from 3 to 9 h resulted in a large increase in the BET surface area, micro pore area and pore diameter from 168.0 m^2/g, 65.2 m^2/g and 18.7 Å to 485.6 m^2/g, 160.9 m^2/g and 36.0 Å, respectively. When the activation time was further increased to 12 h the BET surface area, micro pore area, and pore diameters decreased to 420.5 m^2/g, 149.5 m^2/g and 28.3 Å, respectively (Table 1). By increasing the activation time, the steam widens the pore structure and the surface area increases. When activation is pursued beyond optimum level, the walls of the pores become so thin that they collapse and cause a reduction in the available surface area [27].

Table 1: Effect of activation time on the porous structure of steam-activated carbon produced at temperature of 900 °C and water flow rate of 15 g/h

Activation time (h)	Product yield (wt.%)	BET surface area (m^2/g)	Micropore area (m^2/g)	Average pore diameter (Å)
3	90.5	168.0	65.2	18.7
6	71.3	285.0	146.0	17.1
9	35.1	485.6	160.9	36.0
12	30.2	420.5	149.5	28.3

Effect of Water Flow Rate

The activation of fluid coke was carried out with steam at different water flow rates at a reaction temperature of 900 °C and activation time of 9 h. The outlet gases produced during activation were passed through a cold water bath to recover unreacted steam. This recovered water was measured to estimate the mass of water which reacted with the petroleum coke at the time of activation. Water efficiency was calculated as the percentage of water consumed in activation process and that fed to the reactor during steam activation process. Water efficiency was in the range of 23–38% for experiments performed for 9 h at 900°C for different water flow rates. Higher water efficiency of 38% led to the maximum BET surface area (483 m^2/g) of activated carbon. Table 2 shows the effect of water flow rate on the porous structure of activated carbon product. When the water flow rate was increased from 10.0 to 15.0 g/h, the BET surface area, micropore area, and pore diameters were increased from 359.4 m^2/g, 11.58 m^2/g, 28.1 Å to 482.6 m^2/g,

159.46 m²/g, 37.3 Å, respectively (Table 2). Afterwards, when activation started, the steam developed the pore structure, which increased the surface area [12]. When the water flow rate was increased from 15 to 17.5 g/h, the surface area decreased from 482 to 444 m²/g. This is may be due to the two reasons: pore collapsing at more severe operating conditions and decline in water efficiency from 38 to 24.5% (Table 2). Therefore, water flow rate of 15 g/h was considered as the optimum for further optimization process.

Table 2: Effect of water flow rate on the porous structure of steam-activated carbon at temperature of 900 °C and activation time of 9 h

Water flow rate (g/h)	Water efficiency (%)	Product yield (wt.%)	BET surface area (m^2/g)	Micro pore area (m^2/g)	Average pore diameter (Å)
10.0	35.0	50.1	359.4	11.5	28.1
12.5	36.5	40.3	361.0	87.9	30.2
15.0	38.0	35.2	482.6	159.4	37.3
17.5	24.5	25.4	444.0	114.9	34.7

Effect of Activation Temperature

Table 3 shows the effect of activation temperature on BET surface area, micro pore area and pore diameters of activated carbon produced from fluid coke. When the temperature increased from 600 to 900 °C, the BET surface area, micropore area, and pore diameters of activated carbon also increased from 250 m²/g, 120 m²/g, 25.0 Å to 482 m²/g, 159.0 m²/g, 37.3 Å, which is expected for an overall endothermic process. Increasing the temperature up to 900 °C leads to efficient steam activation due to widening of the porosity and/or the external burning of the carbon particles [22]. However, when the temperature was beyond 900 °C, the surface area, micropore area, and pore diameter decreased to 463 m²/g, 112 m²/g, and 32.4 Å. At this temperature, pore walls become so thin that they collapse and cause a reduction in the available surface area [12]. In addition, the ash content in the product increased with an increase in activation temperature that resulted in lower product yield. Ash offers resistance to the diffusion of the activating agent at higher temperatures and may sinter and block pores as well as reduce the surface area. At optimum steam activation conditions (temperature 900 °C, activation time 9 h, and water flow rate 15 g/h) the surface area, micropore volume and pore diameter were 482 m^2/g, 0.055 cm^3/g and 37.3 Å, respectively.

Table 3: Effect of activation temperature on the porous structure of steam-activated carbon at water flow rate of 15 g/h and activation time of 9 h

Temperature (°C)	Product yield (wt.%)	BET surface area (m^2/g)	Micropore area (m^2/g)	Average pore diameter (Å)
600	85.3	250	120	25.0
700	75.1	350	124	25.1
800	52.3	375	135	30.2
900	35.1	482	159	37.3
1000	25.4	463	112	32.4

Oxidation of Steam-activated Carbons with HNO_3 and H_2O_2

SAC samples were treated with 70% HNO_3 and 30 w/w % H_2O_2 to study the role of surface functional groups on the adsorptive and catalytic properties of activated carbon. As shown in Table 4, surface area decreased from 482 to 310 m^2/g for HNO_3 treatment and to 399 m^2/g for H_2O_2 treatment. The decrease of surface area is mainly due to a decrease in micropore volume, resulting from pore blocking by surface oxide groups produced by these acid treatments. This was confirmed by CHSNO, Boehm analysis, and FTIR results, as discussed later.

Table 4: BET surface area, micropore area, micropore volume, and average pore diameter of precursor and activated carbons prepared from different activation methods

Sample	BET surface area (m^2/g)	Micropore area (m^2/g)	Micropore volume (cm^3/g)	Total pore volume (cm^3/g)	Average pore diameter (Å)
RPC	11.7	1.1	0.001	–	24.6
SAC	482.0	124.1	0.128	0.231	37.3
SAN	310.0	100.3	0.070	0.197	27.8
SAH	399.1	128.4	0.110	0.180	31.6
CAC	38.0	5.3	0.065	0.085	69.0
PAAC	16.2	2.3	0.054	0.068	20.3
KAC	692.4	365.0	0.226	0.264	20.6

H_3PO_4 Activation

PAAC samples showed no significant changes in surface area (16. 2 m^2/g) compared to RPC (11 m^2/g). H_3PO_4 activation is an effective chemical treatment for improving the porous structure of activated carbons when the precursors are lignocellulosic materials. H_3PO_4 activation of biomass is to accelerate the cleavage of bonds between the cellulose and lignin [28]. Fluid coke contains 6 wt.% of sulfur and other metals such as Na, Mg, Al, Ca, Ti, V, Cr, Mn, Fe, Co, Ni, Zn, Sr, Zr, Mo, and Ba (Table 5). These metals can interact with phosphorus and form compound species such as $FePO_4$, $Al(PO_3)_3$, and $Si_2P_2O_7$[28] and [29]. Due to the formation of these compounds, H_3PO_4 activation efficiency decreases with the degree of coalification as evidenced by the decreased surface areas of activated carbons [30]. The metals in ash content of fluid coke can react with phosphoric acid and produce insoluble phosphates that can be trapped inside the carbon matrix due to the cross-linking. Phosphate and polyphosphate species are incorporated into the carbon matrix through C—O—P bonds connecting and cross-linking different organic species that are difficult to remove in the washing step. For these reasons, phosphoric acid treatment did not improve surface area of the activated carbon product.

Table 5: ICP results of raw petroleum coke sample.

Metal impurities determined by ICP in RPC sample																
Name of the metal	Na	Mg	Al	Ca	Ti	V	Cr	Mn	Fe	Co	Ni	Zn	Sr	Zr	Mo	Ba
Metal concentration (ppm)	480	895	9521	931	3082	1406	25	268	9106	14	501	116	108	99	80	101

CO_2 Activation

Fluid-coke is the least reactive carbonaceous material among all types of cokes. Due to their aromatization and graphitization nature, CAC samples did not show any improvement in their porous structure (Table 4). In their study, Furimsky et al. (1998) have shown that after being exposed to the atmosphere for more than 10 years, the oxygen, CO_2, and carbon contents of fluid coke were the same as that of fresh coke [31]. CO_2preferentially reacts in the beginning of CO_2 activation at temperatures up to 300 °C. Thus, CO_2 produced in the reaction zone subsequently reacts with carbon to produce CO, which is the main reducing agent (Eq. (5)).

$$CO_2 + C \rightarrow 2CO \tag{5}$$

KOH Activation

KAC samples showed more surface area (692 m^2/g) and pore volume (0.356 cm^3/g) compared to all other activated carbons prepared in this study (Table 4). In KOH activation, KOH initially transforms into K_2O at high temperature (Eq. (6)) and the resulting K_2O is reduced into metallic potassium (Eq. (7)). After a series of reactions during KOH activation, the end product is K_2CO_3 (Eq. (8)) produced by CO_2 and K_2O to develop a porous structure. K_2CO_3 further reacts with carbon and produces K and CO (Eq. (9)).

$$2KOH \rightarrow K_2O + H_2O \quad (6)$$

$$K_2O + C \rightarrow CO + 2K \quad (7)$$

$$K_2O + CO_2 \rightarrow K_2CO_3 \quad (8)$$

$$K_2CO_3 + 2C \rightarrow 2K + 3CO \quad (9)$$

The reaction of K_2O with carbon as well as K intercalation in pseudographitic layers are two important factors for developing a porous structure in carbon material. Therefore, chemical activation process using KOH reduces the activation time significantly, i.e. in the range of 2 h. This is in agreement with Lee and Choi (2000) and Dipanfilo and Egiebor (1996). SAC, SAN, SAH, and KAC samples had a more significant porous structure than CAC and PAAC samples. Therefore, these four types of activated carbons were used for environmental applications.

ENVIRONMENTAL APPLICATIONS

Four types of activated carbons produced from fluid coke were used for ammonium ion adsorption from aqueous solution and direct catalytic oxidation of H_2S from gas stream.

Application of Activated Carbons for Direct Oxidation of H_2S

The catalytic oxidation of hydrogen sulfide (1 mol% H_2S in CH_4) to sulfur dioxide for gas stream with methane as balance gas was studied at 175 °C, 700 kPa, and oxygen concentration of 1.05 times of stoichiometric ratio (Table 6). The performances were evaluated in terms of breakthrough time (defined at 80% H_2S conversion), g_{Sulfur}/g_{Carbon}, and average production of SO_2. When the H_2S conversion dropped below 80% (2000 ppm for H_2S outlet concentration), the reaction was stopped. g_{Sulfur}/g_{Carbon} shows the amount of sulfur captured per unit mass of activated carbon during the reaction time [24] (Fig. 2). The average production of SO_2 is the ratio of SO_2 produced at various reaction times to the total amount of H_2S fed to the reactor. Table 6 showed that the breakthrough time of 36 min for SAN samples was low compared to that with other samples (SAC, 180 min; SAH, 69 min; KAC, 320 min). This shows that a large surface area in activated carbons provides more storage space for sulfur product that resulted in an increase in the breakthrough time [32].

Table 6: Performance of SAC, SAN, SAH, and KAC samples for direct oxidation of H_2S at reaction temperature of 175 °C, pressure of 700 kPa, and oxygen concentration of 1.05 times of stoichiometric ratio

Sample	Breakthrough time(min)	g_{Sulfur}/ g_{Carbon}	Average SO_2 production(%)	Selectivity(%)
SAC	180	0.62	1.85	99
SAN	36	0.35	0.85	73
SAH	69	1.32	1.25	85
KAC	320	1.75	1.98	99

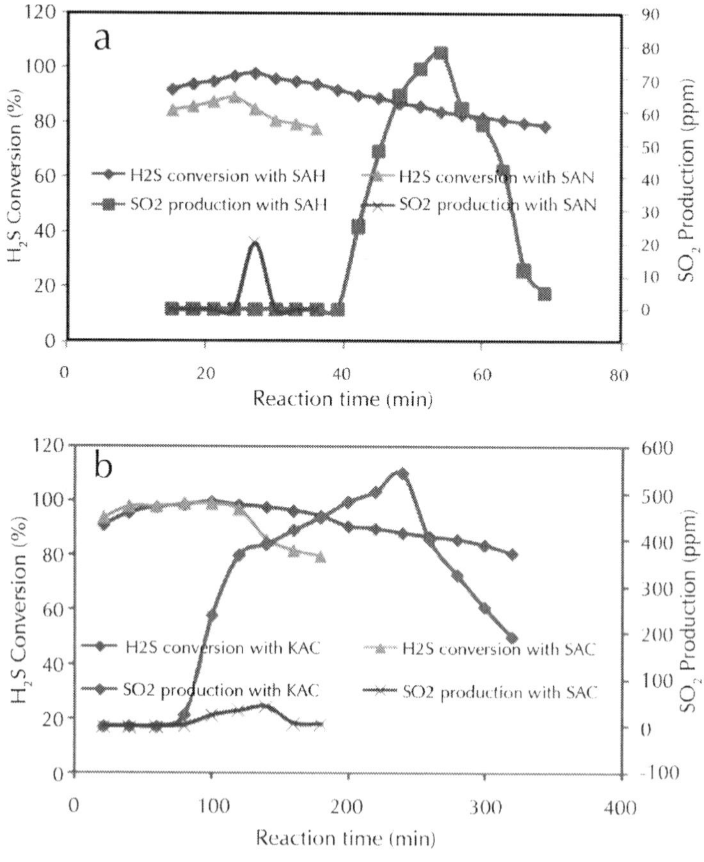

Figure 2: H_2S conversion and SO_2 concentration vs. reaction time for SAC, SAN, SAH, and KAC samples at reaction temperature of 175 °C, pressure of 700 kPa, and oxygen concentration of 1.05 times of stoichiometry.

Hydrogen sulfide oxidation and SO_2 production can be shown using the following reactions:

$$H_2S + \left(\frac{1}{2}\right)O_2 \rightarrow S + H_2O \quad \Delta H = -206 kJ/mol \tag{10}$$

$$S + O_2 \rightarrow SO_2 \quad \Delta H = -297 kJ/mol \tag{11}$$

In Table 6, SAN showed the smallest breakthrough time for H_2S oxidation. It can be due to the large amount of carboxylic groups (see Table 7) and its acidic nature, which are not suitable for adsorption of H_2S on the surface of activated carbon as the first step, required for the oxidation reaction. In addition, it has the lowest BET surface area (see Table 4) among four types of activated carbons used for this study. This cannot provide large storage space for sulfur produced by reaction and resulted in fast deactivation of catalyst. In case of KAC, its basic character (see Table 7) and large total pore volume (0.264 cm^3/g) help in enhancing its performance and large value for g_{Sulfur}/g_{Carbon}. SAC showed better performance compared with that of SAH, which is due to SAC better porous characteristics (Table 4). SO_2 is a pollutant and an undesirable by-product. Therefore, its production during H_2S oxidation reaction should be kept as low as possible. SO_2 is produced by the surface reaction between sulfur and oxygen on the carbon surface (Eq. (11)). For all activated carbon samples, SO_2 production increased after dropping H_2S conversion from the maximum value. At this condition, due to the large amount of sulfur deposited on carbon surface, there are higher chances of reaction between oxygen and sulfur on carbon surface. Then, when there are larger amount of unreacted H_2S, the concentration of SO_2 decreases by the following reaction.

$$SO_2 + 2H_2S \rightarrow 3S + 2H_2O \qquad \Delta H = -146 \text{ kJ/mol} \quad (12)$$

Table 7: Boehm titration results for original coke and prepared activated carbons.

Acidic and basic functional group concentration(mmol/g)						
Sample	**Carbox-ylic**	**Lactonic**	**Phenolic**	**Acidic**	**Basic**	**Total**
RAC	0.125	1.125	0.375	1.625	0.851	2.477
SAC	0.015	0.377	0.375	0.752	0.375	1.127
SAN	1.523	0.375	0.512	2.410	0.500	2.910
SAH	0.375	0.125	0.625	1.125	0.750	1.875
KAC	0.010	0.115	0.215	0.154	1.752	2.256

Application for Ammonium Ion Adsorption

Batch experiments were carried out in 125 cm^3 glass capped vials containing 1.25 g of adsorbent and 125.0 ml of two different concentrations (40 mg/l and 260 mg/l) of ammonium hydroxide solution to examine the adsorption efficiency of SAC, SAN, SAH, and KAC samples at room temperature (26 °C) and at pH 8.0. The pH of the aqueous phase is an important parameter for adsorption of ammonium ion. According to the Marañón et al. (2006) and Kadir et al. (2007) the removal efficiency of ammonium ion increased by raising the pH from 4 to 8 and reached a maximum value at pH 8, beyond which it decreased. The decrease in the NH_4^+ removal efficiency after pH 8 may be due to the conversion of NH_4^+ into NH_3[33] and [34]. In the present study related to ammonium ion adsorption experiments, pH was adjusted to 8.0 with 0.025 N NaOH and 0.025 N H_2SO_4 for the maximum adsorption of NH_4^+. The experimental results in Fig. 3 show that the ammonium ion adsorption *qe* (mg/g) on SAC, SAN, SAH, and KAC activated carbons increased from 28.9, 37.7, 32.8, and 8.5 to 82.6, 184.6, 169.6, and 66.7 when the ammonium ion concentration increased from 40 mg/l to 260 mg/l. The equilibrium time for SAC, SAN, SAH and KAC is respectively 7 h, 3 h, 4 h and 5 h. This is because of the mass-transfer difference for different samples, between the liquid and the surface of activated carbon. When the concentration of adsorbate increased, the adsorption of ammonium ion increased due to diffusion of ammonium ions that also increased. In addition, the ammonium ions on the surface of activated carbon samples in equilibrium with the ammonium ions in the solution increased with its concentration in the solution. These results are agreement with those of other researchers [35].

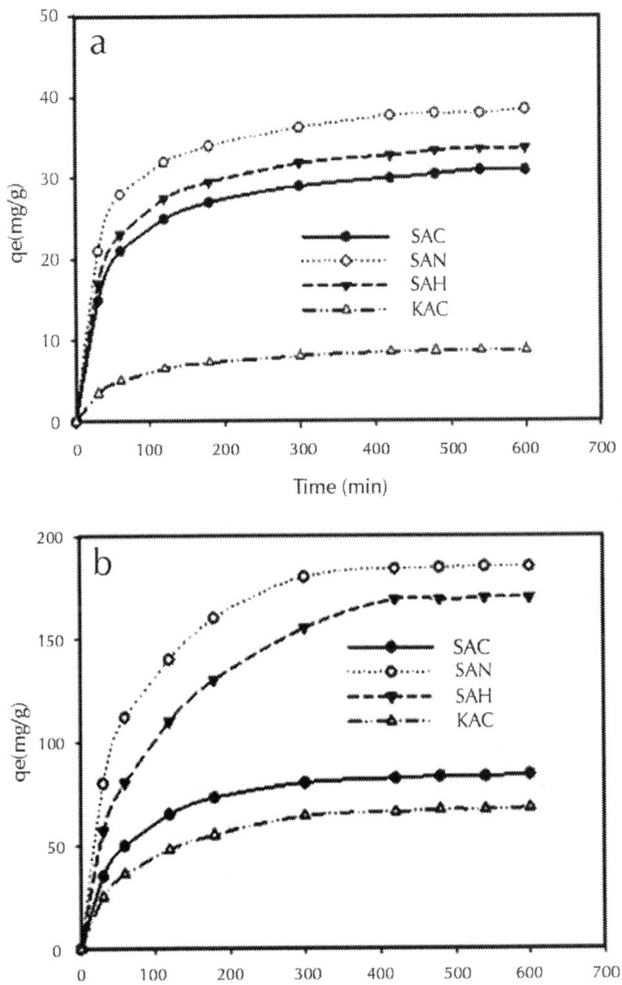

Figure 3: Effect of initial concentration (a) 40 mg/l (b) 260 mg/l and contact time on ammonium ion adsorption.

Oxidized samples such as SAN and SAH showed nearly two times more ammonium ion removal efficiency compared to that for SAC and KAC samples. Table 4 shows that SAC and KAC samples had larger surface area, and pore volume, but less adsorption capacities than the SAN and SAH samples. On the basis of structural properties of KAC and SAC samples, it may be concluded that the ammonium adsorption does not depend only on porous structure of carbon samples, but may also depend on acid–basic interactions between the ammonium ion and the acid surface functional groups of the adsorbents [36]. Boehm results (Table 7) showed that the total amount of surface functional groups for SAC and KAC was less than that for SAN and SAH samples. SAN samples showed more adsorption capacity than SAH samples due to more oxygenated groups in SAN samples. This means that at higher concentrations, ammonium ions participate in the formation of hydrogen bonds, competing with the water molecules, with O-containing groups or electron donating groups present on the carbon surface [36].

POROUS CHARACTERIZATION AND SURFACE CHEMISTRY OF ACTIVATED CARBONS

The porous structure of activated carbon (micro, meso and macro pores) is developed as a function of operating conditions used for the activation process. Table 8 shows the ultramicropore (≤ 7 Å), supermicropore (7–20 Å), micropore, mesopore, total pore volume, and BET surface area of SAC, SAN, SAH, and KAC activated carbons. All parameters, except BET surface area, are measured using the density functional theory (DFT) method. For activated carbons prepared by steam activation, porous characteristics including surface area, micropore area, pore volume, and average pore diameter increased to 482 m^2/g, 124.1 m^2/g, 0.231 cm^3/g, and 37.3 Å, respectively (Table 4).

Table 8: Porous characteristics of petroleum coke and activated carbons prepared by different activation methods at optimum operating conditions (measured using DFT)

Sample	Pore volume for pores (< 7 Å) cm^3/g	Pore volume for pores (< 7–20 Å) cm^3/g	Micropore volume for pores (< 20 Å) cm^3/g	Mesopore volume for pores (20–500 Å) cm^3/g	Total pore volume (cm^3/g)	BET surface area (m^2/g)
SAC	0.060	0.067	0.128	0.104	0.231	482
SAN	0.040	0.029	0.070	0.122	0.197	310
SAH	0.049	0.060	0.110	0.069	0.180	399
KAC	0.119	0.107	0.226	0.034	0.264	692

Fig. 4(a) and (b) shows the incremental micropore volume and the mesopore volume, respectively for SAC, SAN, SAH, and KAC samples. KAC and SAC samples showed better performance for H_2S oxidation than SAH and SAN samples. This can be explained by H_2S having a small molecular size (1.336 Å), suggesting that pores with diameter less than 7 Å are favorable for their physical adsorption due to the enhancement of adsorption potential. KAC samples are highly porous, with a narrow range of pore size (ultramicropore: < 7 Å, 0.119 cm^3/g) and high surface area (692 m^2/g). SAC samples exhibited selectively towards H_2S oxidation on the basis of their pore sizes (0.060 cm^3/g of ultramicropore and 482 m^2/g surface area). Surface chemistry of carbons also plays a significant role for H_2S adsorption/oxidation, due to its acidic nature [37]. Boehm results showed (Table 7) more basic groups (1.752 mmol/g) for KAC than other samples.

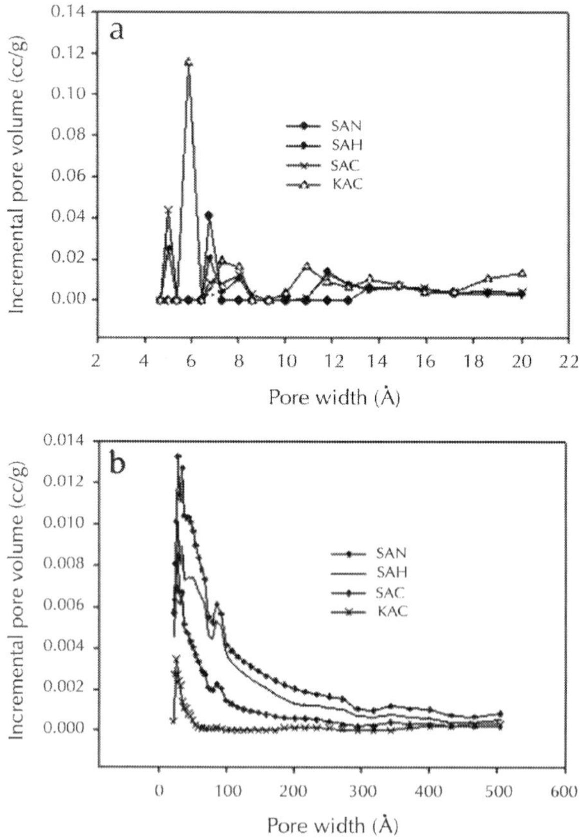

Figure 4: Incremental micropore volume (a) and mesopore volume (b) for different activated carbons (measured using DFT).

SAN and SAH samples have lower incremental pore volume for the entire range of micropores and mesopores than SAC and KAC samples. This is because HNO_3 and H_2O_2 treatments blocked some pores by the introduction of surface functional groups, resulting in a decrease in pore volume. For SAN samples, ultramicropore (< 7 Å) volume, micropore (< 20 Å) volume, and total pore volume decreased to 0.040 cm^3/g, 0.070 cm^3/g, and 0.197 cm^3/g, while SAH samples had ultramicropore volume of 0.049 cm^3/g, micropore volume of 0.110 cm^3/g, and a total pore volume of 0.180 cm^3/g. This can be explained by more severe reactions of HNO_3 with carbon material and the introduction of functional groups, more with HNO_3 treatment than that for H_2O_2 treatment, as observed by Boehm, FTIR, and elemental analyses which are discussed in the following sections.

Table 9 shows the elemental analysis of activated carbon samples obtained by different activation methods. Carbon reactions with HNO_3 and H_2O_2 have been postulated to involve electron donations or acceptance mechanisms, as well as the free radical intermediates (hydroxyl OH^- and peroxide OOH^-). This shows the importance of chemical functional groups on the activated carbon surface. However, this oxidation treatment gives rise to a large increase in total acidity, resulting from the increase in surface oxide groups such as carboxyl, lactone, and phenoic groups.

Table 9: Elemental analysis of RPC and activated carbons

Samples	N (wt.%)	C (wt.%)	S (wt.%)	H (wt.%)	Ash (wt.%)	O[a] (wt.%)
RPC	1.76	76.25	6.48	1.47	8.50	5.54
SAC	0.74	73.11	4.22	0.38	14.94	6.61
SAN	1.98	72.37	3.67	0.27	9.43	12.28
SAH	0.62	72.82	2.86	0.34	14.02	9.34
CAC	0.87	76.50	6.24	0.41	12.26	3.72
PAAC	0.55	80.35	6.16	0.48	9.13	3.33
KAC	1.25	84.94	0.22	0.55	10.33	2.71

[a]Oxygen is determined by difference.

The increment of oxygen content in SAN and SAH samples compared with that for SAC is due to the formation of surface oxygen functional groups on the activated carbon surface by HNO_3 and H_2O_2treatments. The increase in nitrogen content in SAN samples was due to the incorporation of nitrogen functional groups such as N–O or $CONH_2$ during the HNO_3 treatment of activated carbons [38]. In SAC, CAC, PAAC and KAC samples, ash contents were increased compared with that for RPC due to the weight loss occurred in the former samples because of the higher removal of unstable carbons during the activation process. In the case of SAN and SAH samples, ash contents were also lower than SAC due to the removal of ash by HNO_3 and H_2O_2 treatments.

Fig. 5 presents the FTIR spectra of raw petroleum coke as well as obtained activated carbons. RPC sample showed the adsorption bands at 3637 and 3024 cm^{-1} which are O—H stretching in alcohols and C—H stretching in alkanes respectively. Other absorption bands at 2932, 2910, and 1380 cm^{-1} can be assigned to asymmetric and symmetric —C—H stretching vibrations in aliphatic such as $—CH_2CH_3$, $=CH_3$, and $—CH_3$. The bands around 1610 cm^{-1} and 1442 cm^{-1} are assigned to the —C=C— aromatic bonds and exalted by the presence of phenolic groups. The broad band at approximately 1150 to 1200 cm^{-1} may be due to the overlapping of C—O—O stretching, C—O stretching and O—H bending modes of alcoholic, phenolic, and carboxylic groups [39]. The adsorption peaks at 700–900 cm^{-1} represent aromatic C—H bands.

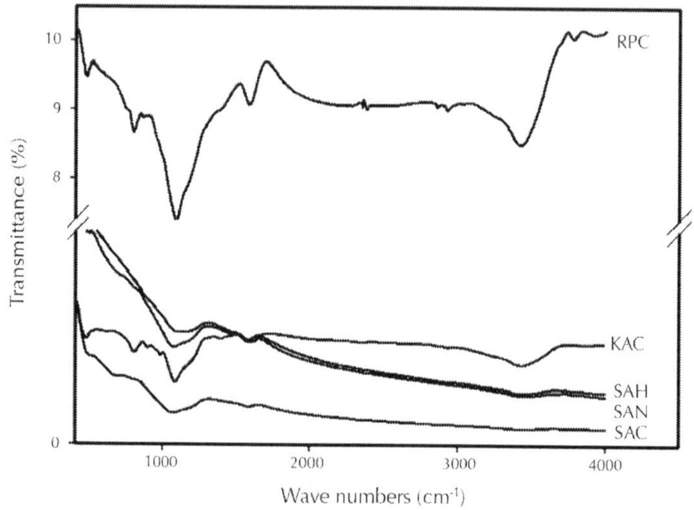

Figure 5: FTIR spectra for SAC, SAN, SAH, and KAC activated carbons.

The SAC sample absorption bands at 1200–1220 cm^{-1} are ascribable to stretching C—O vibrations. The other band at 1720 and 1650 cm^{-1} may be developed by CO vibrations in oxygen surface groups. However, IR band intensities decreased in the SAC sample compared to the RPC sample, indicating a reduction in the number of the IR-observable species. SAC samples obtained at 900 °C significantly removed all the surface groups of the RPC sample, as shown in the literature [40] and [41]. At a lower temperature range (120–500 °C) strong acid groups (carboxylic groups) decompose easier than weak acid groups (phenols and carbonyls) and oxygen-containing organic bases (500–800 °C). The oxygen surface functionalities at higher temperatures decompose to water, carbon monoxide, and carbon dioxide. CO_2evolution is related to carboxylic groups, some anhydride and lactone groups, and less acidic groups containing —OH functions [42] and [43]. These results are in agreement with the Boehm analysis, which showed that the total surface groups decreased from 2.477 mmol/g in the RPC samples to 1.127 mmol/g in the SAC samples, due to high activation temperature of 900 °C (Table 7).

The treatment of SAC by both HNO_3 and H_2O_2 greatly appears on the absorption bands. This implies that the HNO_3 and H_2O_2 treatment processes increased the number of strong and weak acidic functional groups, respectively. HNO_3 treatment increases the quantity of acidic surface functional groups [44]. Various surface oxygen groups and structures containing N—O bonds (nitro groups and nitrate complexes) are developed [45]. It is worth noticing that the weak band appearing at 1710 cm^{-1} and the aromatic C═C vibration band at 1600 cm^{-1} changed into large peaks after HNO_3 and H_2O_2 treatment. The bands in the region 700–900 cm^{-1} are assigned to the out-of-plane vibration of the isolated and adjacent aromatic C—H groups. The C—O and C—O—C bands at the wavelengths of 1200–1300 cm^{-1} also represent etheric oxygen groups. From these results, it can be concluded that H_2O_2 is a less effective oxidant than HNO_3, concerning the number of surface functional groups that improved. However, it also clearly indicates that the oxidation by HNO_3 and H_2O_2 generates a large number of surface functional groups such as nitrate, carbonyl and carboxyl groups. These results correlate with the elemental analysis (Table 9) and the Boehm results (Table 7), showing an increase in surface functional groups from 1.127 mmol/g for the SAC samples to 2.91 mmol/g for the SAN samples, and 1.875 mmol/g for the SAH samples.

In the KAC samples, adsorption bands in the region of 700–900 cm^{-1} are observed adjacent to aromatic C—H groups. The peaks at 1600 cm^{-1} indicate that the C=C bands are due to the vibration of oxygen groups enhanced by ether in the carbon matrices. The C—O and C—O—C bands at 1200–1300 cm^{-1}also represent oxygen groups. In the KAC sample, the major functional groups are C—O, C—O—C, and—OH groups. These results correlate with the Boehm results (Table 7), showing that basic groups increased from 0.851 mmol/g in the RPC sample to 1.752 mmol/g in the KAC sample.

The BET surface area analysis revealed that the SAC and KAC samples had a more developed porous structure compared to that for the SAN, SAH, CAC, and PAAC samples, as shown in the scanning electron micrographs in Fig. 6. According to the SEM micrographs, raw petroleum coke has undeveloped porous structure (Fig. 6(a)), which is also confirmed by its low BET surface area (11 m^2/g). This first observation represents a clear attack of its surface. Fig. 6(b) shows how the steam affects the whole material through deeper cracks appearing on the grain. Under optimum steam activation conditions (temperature 900 °C, water flow rate 15 g/h, and 9 h activation time), a deep and general fissuration of the whole grain was evident. Fissures started to extend more profoundly, while the pre-existing concentric layers of the fluid coke particle peeled off, resulting in a surface area increase to 482 m^2/g (Fig. 6(c)). The KAC sample inFig. 6(d) shows mesoporosity, which is typical for activation by KOH, due to the formation of intercalation in pseudographitic layers and the shrinkage property of coke, which helps to form a lamella like structure. In effect, the coke swelled with the amount of volatile matter when the KOH reacted with the petroleum coke.

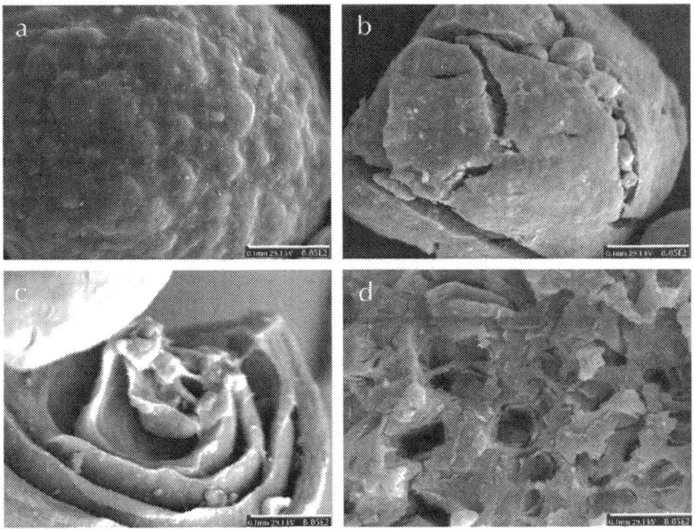

Figure 6: Scanning electron micrographs of raw coke and activated carbons (a), RPC (b), SAC at 500 °C, (c) SAC at 900 °C, and (d) KAC [Magnification: 1490 ×].

CONCLUSIONS

Activated carbon was produced from fluid petroleum coke using physical and chemical activation methods and characterized by BET, FTIR, Boehm, CHSNO, ICPMS and SEM analyses. Steam and KOH activation methods improved the BET surface area and pore volume (482 m^2/g, 0.231 cm^3/g and 692 m^2/g 0.264 cm^3/g, respectively) compared to those by CO_2 and H_3PO_4 activation methods. The oxidation of SAC samples with HNO_3 (SAN) and H_2O_2 (SAH) showed more surface functional groups and decreased the surface area ~ 15%, which was confirmed by FTIR, Boehm, and BET analyses. The order of ammonium ion removal from aqueous solutions was SAN > SAH > SAC > KAC. Due to the oxygenated functional groups, SAN and SAH samples showed more ammonium adsorption than other samples. The order of H_2S breakthrough time of direct catalytic oxidation showed KAC > SAC > SAH > SAN, due to the porous structure and surface functionalities of activated carbons. It was concluded that activated carbons produced from fluid petroleum coke using steam and KOH activation methods are more suitable for H_2S oxidation and that HNO_3 and H_2O_2 treated samples were more suitable for ammonium ion adsorption.

ACKNOWLEDGMENT

The authors greatly acknowledge the Syncrude Canada Limited and the Natural Sciences and Engineering Research Council of Canada (NSERC) for financial support and Drs. S. Khalid and K. Pasha from the Department of Chemical Technology, SIAST (Canada) for providing Calorimeter (MC-500) for this work.

REFERENCES

1. R.W. Bryers, Utilization of petroleum coke and petroleum coke/coal blends as a means of steam raising, Fuel Processing Technology 44 (1995) 121–141.
2. N. Berguerand, A. Lyngfelt, The use of petroleum coke as fuel in a 10 kWth chemical-looping combustor, International Journal of Greenhouse Gas Control 2 (2008) 169–179.
3. D. Trommer, F. Noembrini, M. Fasciana, D. Rodriguez, A. Morales, M. Romero, et al., Hydrogen production by steam-gasification of petroleum coke using concentrated solar power — I. Thermodynamic and kinetic analyses, International Journal of Hydrogen Energy 30 (2005) 605–618.
4. T. Otowa, Y. Nojima, T. Miyazaki, Development of KOH activated high surface area carbon and its application to drinking water purification, Carbon 35 (1997) 1315–1319.
5. J. Wang, E.J. Anthony, J.C. Abanades, Clean and efficient use of petroleum coke for combustion and power generation, Fuel 83 (2004) 1341–1348.

6. H. Leion, T. Mattisson, A. Lyngfelt, The use of petroleum coke as fuel in chemical-l ooping combustion, Fuel 86 (2007) 1947–1958.
7. S.H. Lee, C.S. Choi, Chemical activation of high sulfur petroleum cokes by alkali metal compounds, Fuel Processing Technology 64 (2000) 141–153.
8. C. Lu, S. Xu, C. Liu, The role of K2CO3 during the chemical activation of petroleum coke with KOH, Journal of Analytical and Applied Pyrolysis 87 (2010) 282–287.
9. M. Wu, Q. Zha, J. Qiu, X. Han, Y. Guo, Z. Li, et al., Preparation of porous carbons from petroleum coke by different activation methods, Fuel 84 (2005) 1992–1997.
10. T. Kawano, M. Kubota, M.S. Onyango, F. Watanabe, H. Matsuda, Preparation of activated carbon from petroleum coke by KOH chemical activation for adsorption heat pump, Applied Thermal Engineering 28 (2008) 865–871.
11. B. Jiang, Y. Zhang, J. Zhou, K. Zhang, S. Chen, Effects of chemical modification of petroleum cokes on the properties of the resulting activated carbon, Fuel 87 (2008) 1844–1848.
12. R. DiPanfilo, N.O. Egiebor, Activated carbon production from synthetic crude coke, Fuel Processing Technology 46 (1996) 157–169.
13. C. Saha, E.L. Tollefson, Hydrodesulfurization of athabasca fluid coke — conversion and mechanism, Canadian Journal of Chemical Engineering 73 (2) (1995) 211–219.
14. http://www.prlog.org/11537927-petroleum-coke-industry-to-2015.html.
15. E.W. Dennis, G.J. Green, A laboratory study of petroleum coke combustion: kinetics and catalytic effects, Industrial and Engineering Chemistry Research 27 (7) (1988) 1115–1120.
16. J.J. McKetta, Petroleum Processing Handbook, Marcel Dekker, New York, 1992.
17. B.S. Parmar, E.L. Tollefson, Desulfurization of oil sands coke, Canadian Journal of Chemical Engineering 55 (2) (1997) 185–191.

18. P. Chingombe, B. Saha, R.J. Wakeman, Surface modification and characterisation of a coal-based activated carbon, Carbon 43 (2005) 3132–3143.
19. http://stocks.investopedia.com/stock-analysis/2011/Canadian-Oil-ProductionSet-To-Grow-Rapidly-RDS-MRO-CVX-NXY-SU-MUR-IMO-SNP-CVE0607.aspx#axzz1XyAMg6KD.
20. http://www.atsdr.cdc.gov/toxguides/toxguide-114.pdf?id=1088&tid=67.
21. http://www.atsdr.cdc.gov/ToxProfiles/tp126-c1.pdf.
22. R. Azargohar, A.K. Dalai, Steam and KOH activation of biochar: experimental and modeling studies, Microporous and Mesoporous Materials 110 (2008) 413–421.
23. H.P. Boehm, Surface oxides on carbon and their analysis: a critical assessment, Carbon 40 (2002) 145–149.
24. R. Azargohar, A.K. Dalai, The direct oxidation of hydrogen sulphide over activated carbons prepared from lignite coal and biochar, Canadian Journal of Chemical Engineering 89 (2011) 844–853.
25. H. Muhlen, K.H. Van Heek, in: W.J. Patrick (Ed.), Porosity in Carbons: Characterization and Applications, Halsted Press, London, UK, 1995, Chapter 5.
26. F. Rodríguez-Reinoso, M. Molina-Sabio, Textural and chemical characterization of microporous carbons, Advances in Colloid and Interface Science 76–77 (1998) 271–294.
27. Y. Huang, B. Jin, Z. Zhong, W. Zhong, R. Xiao, Characteristic and mercury adsorption of activated carbon produced by CO2 of chicken waste, Journal of Environmental Sciences 20 (2008) 291–296.
28. M. Jagtoyen, M. Thwaites, J. Stencel, B. McEnaney, F. Derbyshire, Adsorbent carbon synthesis from coals by phosphoric acid activation, Carbon 30 (1992) 1089–1096.
29. F. Rodríguez-Reinoso, M. Molina-Sabio, Activated carbons from lignocellulosic materials by chemical and/or physical activation: an overview, Carbon 30 (1992) 1111–1118.

30. M. Jagtoyen, F. Derbyshire, Some considerations of the origins of porosity in carbons from chemically activated wood, Carbon 31 (1993) 1185–1192.

31. E. Furimsky, Gasification of oil sand coke: review, Fuel Processing Technology 56 (1998) 263–290.

32. H. Marsh, F.R. Reinoso, Activated Carbon, 978-0-08-044463-5, 2006.

33. E. Marañón, M. Ulmanu, Y. Fernández, I. Anger, L. Castrillón, Removal of ammonium from aqueous solutions with volcanic tuff, Journal of Hazardous Materials 137 (2006) 1402–1409.

34. K. Saltalı, A. Sarı, M. Aydın, Removal of ammonium ion from aqueous solution by natural Turkish (Yıldızeli) zeolite for environmental quality, Journal of Hazardous Materials 141 (2007) 258–263.

35. M.A. Wahab, S. Jellali, N. Jedidi, Ammonium biosorption onto sawdust: FTIR analysis, kinetics and adsorption isotherms modeling, Bioresource Technology 101 (2010) 5070–5075.

36. P. Vassileva, P. Tzvetkova, R. Nickolov, Removal of ammonium ions from aqueous solutions with coal-based activated carbons modified by oxidation, Fuel 88 (2009) 387–390.

37. M. Steijns, P. Mars, Catalytic oxidation of hydrogen sulfide. influence of pore structure and chemical composition of various porous substances, Industrial & Engineering Chemistry Product Research and Development 16 (1977) 35–41.

38. J.P. Chen, S. Wu, Acid/Base-treated activated carbons: characterization of functional groups and metal adsorptive properties, Langmuir 20 (2004) 2233–2242.

39. L. Chunlan, X. Shaoping, G. Yixiong, L. Shuqin, L. Changhou, Effect of precarbonization of petroleum cokes on chemical activation process with KOH, Carbon 43 (2005) 2295–2301.

40. V. Gomez-Serrano, J. Pastor-Villegas, A. Perez-Florindo, C. Duran-Valle, C. Valenzuela-Calahorro, FT-IR study of rockrose and of char and activated carbon, Journal of Analytical and Applied Pyrolysis 36 (1996) 71–80.

41. D.V. Brazhnyk, Y.P. Zaitsev, I.V. Bacherikova, V.A. Zazhigalov, J. Stoch, A. Kowal, Oxidation of H2S on activated carbon KAU and influence of the surface state, Applied Catalysis B: Environmental 70 (2007) 557–566.
42. Y. Otake, R.G. Jenkins, Characterization of oxygen-containing surface complexes created on a microporous carbon by air and nitric acid treatment, Carbon 31 (1993) 109–121.
43. M. Molina-Sabio, A. Munecas, F. Rodriguez-Reinoso, Characterization of Porous Solids II, First ed. Elsevier, Holland, 1991.
44. J.S. Noh, J.A. Schwarz, Effect of HNO3 treatment on the surface acidity of activated carbons, Carbon 28 (1990) 675–682.
45. V. Gómez-Serrano, M. Acedo-Ramos, A.J. López-Peinado, C. Valenzuela- Calahorro, Mass and surface changes of activated carbon treated with nitric acid. Thermal behavior of the samples, Thermochimica Acta 291 (1997) 109–115.

Citations

CHAPTER 1

Sachin Kumar and R. K. Singh, "Thermolysis of High-Density Polyethylene to Petroleum Products," Journal of Petroleum Engineering, vol. 2013, Article ID 987568, 7 pages, 2013. doi:10.1155/2013/987568.

CHAPTER 2

Abdulaziz, A. (2014) Evaluation of Microseismicity Related to Hydraulic Fracking Operations of Petroleum Reservoirs and Its

Possible Environmental Repercussions. Open Journal of Earthquake Research, 3, 43-54. doi:10.4236/ojer.2014.32006.

CHAPTER 3

Feyzullayev, A. and Ibragimov, V. (2014) Environmental Consequences of Long-Term Development of Petroleum Fields, Absheron p-la, Azerbaijan, Case History. Journal of Environmental Protection, 5, 1603-1610. doi: 10.4236/jep.2014.517151.

CHAPTER 4

Nilanjana Das and Preethy Chandran, "Microbial Degradation of Petroleum Hydrocarbon Contaminants: An Overview," Biotechnology Research International, vol. 2011, Article ID 941810, 13 pages, 2011. doi:10.4061/2011/941810.

CHAPTER 5

Kim, J., Min, K. and Yim, E. (2014) Experimental Investigation of Fuel Quality and Contaminant Materials from Liquefied Petroleum Fuel. Engineering, 6, 644-653. Doi: 10.4236/eng.2014.610064.

CHAPTER 6

O. Oluwole and E. Emagbetere, "Finite Element Analysis of In-Plane Displacements and Von-Mises Stresses in Ellipsoidal and Circular Cylinderical Petroleum Tankers," Engineering, Vol. 5 No. 2, 2013, pp. 167-177. doi:10.4236/eng.2013.52024.

CHAPTER 7

Yuan Yao, Diane Graziano, Matthew Riddle, Joe Cresko, Eric Masanet, Greener pathways for energy-intensive commodity chemicals: opportunities and challenges, Current Opinion in Chemical Engineering, Volume 6, November 2014, Pages 90-98, ISSN 2211-3398, http://dx.doi.org/10.1016/j.coche.2014.10.005.

CHAPTER 8

B. Bennett, A. Lager, D.K. Potter, J.O. Buckman, S.R. Larter, Petroleum geochemical proxies for reservoir engineering parameters, Journal of Petroleum Science and Engineering, Volume 58, Issues 3–4, September 2007, Pages 355-366, ISSN 0920-4105, http://dx.doi.org/10.1016/j.petrol.2006.06.009.

CHAPTER 9

N. Rambabu, R. Azargohar, A.K. Dalai, J. Adjaye, Evaluation and comparison of enrichment efficiency of physical/chemical activations and functionalized activated carbons derived from fluid petroleum coke for environmental applications, Fuel Processing Technology, Volume 106, February 2013, Pages 501-510, ISSN 0378-3820, http://dx.doi.org/10.1016/j.fuproc.2012.09.019.

Index